Stealth Assessment

M000047472

This report was made possible by grants from the John D. and Catherine T. MacArthur Foundation in connection with its grant making initiative on Digital Media and Learning. For more information on the initiative visit http://www.macfound.org.

The John D. and Catherine T. MacArthur Foundation Reports on Digital Media and Learning

Peer Participation and Software: What Mozilla Has to Teach Government, by David R. Booth

Kids and Credibility: An Empirical Examination of Youth, Digital Media Use, and Information Credibility, by Andrew J. Flanagin and Miriam Metzger with Ethan Hartsell, Alex Markov, Ryan Medders, Rebekah Pure, and Elisia Choi

New Digital Media and Learning as an Emerging Area and "Worked Examples" as One Way Forward, by James Paul Gee

Digital Media and Technology in Afterschool Programs, Libraries, and Museums, by Becky Herr-Stephenson, Diana Rhoten, Dan Perkel, and Christo Sims with contributions from Anne Balsamo, Maura Klosterman, and Susana Smith Bautista

Quest to Learn: Developing the School for Digital Kids, by Katie Salen, Robert Torres, Loretta Wolozin, Rebecca Rufo-Tepper, and Arana Shapiro

Measuring What Matters Most: Choice-Based Assessments for the Digital Age, by Daniel L. Schwartz and Dylan Arena

Learning at Not-School? A Review of Study, Theory, and Advocacy for Education in Non-Formal Settings, by Julian Sefton-Green

Stealth Assessment: Measuring and Supporting Learning in Video Games, by Valerie Shute and Matthew Ventura

The Future of the Curriculum: School Knowledge in the Digital Age, by Ben Williamson

For a complete list of titles in this series, see http://mitpress.mit.edu/books/series/john-d-and-catherine-t-macarthur-foundation-reports-digital-media-and-learning.

Stealth Assessment

Measuring and Supporting Learning in Video Games

Valerie Shute and Matthew Ventura

The MIT Press
Cambridge, Massachusetts
London, England

MIT Press books may be purchased at special quantity discounts for business or sales promotional use. For information, please email special_sales@mitpress.mit.edu or write to Special Sales Department, The MIT Press, 55 Hayward Street, Cambridge, MA 02142.

This book was set in Stone Serif and Stone Sans by the MIT Press. Printed and bound in the United States of America.

Library of Congress Cataloging-in-Publication Data
Shute, Valerie J. (Valerie Jean), 1953– , author.
Stealth assessment : measuring and supporting learning in video games / Valerie Shute and Matthew Ventura.
   pages   cm. — (The John D. and Catherine T. MacArthur Foundation reports on digital media and learning)
Includes bibliographical references.
ISBN 978-0-262-51881-9 (pbk. : alk. paper)
1. Educational tests and measurements. 2. Video games. I. Ventura, Matthew, author. II. Title.
LB3051.S518   2013
371.26—dc23
2012038217

10   9   8   7   6   5   4   3   2   1

Contents

Series Foreword

The John D. and Catherine T. MacArthur Foundation Reports on Digital Media and Learning, published by the MIT Press in collaboration with the Monterey Institute for Technology and Education (MITE), present findings from current research on how young people learn, play, socialize, and participate in civic life. The reports result from research projects funded by the Mac-Arthur Foundation as part of its fifty million dollar initiative in digital media and learning. They are published openly online (as well as in print) in order to support broad dissemination and stimulate further research in the field.

Acknowledgments

We would like to sincerely thank the Bill and Melinda Gates Foundation for its funding for this project, particularly Emily Dalton-Smith, Robert Torres, and Ed Dieterle. We would also like to express our appreciation to the other members of the research grant team—Yoon Jeon Kim, Don Franceschetti, Russell Almond, Matt Small, and Lubin Wang—for their awesome and abundant support on the project, and Lance King, who came up with the "agents of force and motion" idea. Finally, we acknowledge Diego Zapata-Rivera for ongoing substantive conversations with us on the topic of stealth assessment.

Education in the Twenty-First Century

You can discover more about a person in an hour of play than in a year of conversation.

—Plato

In the first half of the twentieth century, a person who acquired basic reading, writing, and math skills was considered to be sufficiently literate to enter the work force (Kliebard 1987). The goal back then was to prepare young people as service workers, because 90 percent of the students were not expected to seek or hold professional careers (see Shute 2007). With the emergence of the Internet, however, the world has become more interconnected, effectively smaller, and more complex than before (Friedman 2005). Developed countries now rely on their knowledge workers to deal with an array of complex problems, many with global ramifications (e.g., climate change or renewable energy sources). When confronted by such problems, tomorrow's workers need to be able to think systemically, creatively, and critically (see, e.g., Shute and Torres 2012; Walberg and Stariha 1992).

These skills are a few of what many educators are calling twenty-first-century (or complex) competencies (see Partnership for the 21st Century 2012; Trilling and Fadel 2009).

Preparing K–16 students to succeed in the twenty-first century requires fresh thinking about what knowledge and skills (i.e., what we call competencies) should be taught in our nation's schools. In addition, there's a need to design and develop valid assessments to measure and support these competencies. Except in rare instances, our current education system neither teaches nor assesses these new competencies despite a growing body of research showing that competencies, such as persistence, creativity, self-efficacy, openness, and teamwork (to name a few), can substantially impact student academic achievement (Noftle and Robins 2007; O'Connor and Paunonen 2007; Poropat 2009; Sternberg 2006; Trapmann et al. 2007). Furthermore, the methods of assessment are often too simplified, abstract, and decontextualized to suit current education needs. Our current assessments in many cases fail to assess what students actually can do with the knowledge and skills learned in school (Shute 2009). What we need are new performance-based assessments that assess how students use knowledge and skills that are directly relevant for use in the real world.

One challenge with developing a performance-based measure is crafting appropriate situations or problems to elicit a competency of interest. A way to approach this problem is to use digital learning environments to simulate problems for performance-based assessment (Dede 2005; DiCerbo and Behrens 2012; Quellmalz et al. 2012). Digital learning environments can provide meaningful assessment environments by supplying students with scenarios that require the application of various competencies. This report introduces a variant of this assessment approach by investigating how performance-based assessments can be used in digital games. Specifically, we are interested in how assessment in games can be used to enhance learning (i.e., formative assessment).

For example, consider role-playing games (e.g., *World of Warcraft*). In these games, players must read lengthy and complex quest logs that tell them the goals. Without comprehending these quest instructions, the players would not be able to know how to proceed and succeed in the game. This seemingly simple task in role-playing games is, in fact, an authentic, situated assessment of reading comprehension. Without these situated and meaningful assessments, we cannot determine what students can actually do with the skills and knowledge obtained. Thus new, embedded, authentic types of assessment methods are needed to properly assess valued competencies.

Why use well-designed games as vehicles to assess and support learning? There are several reasons. First, as our schools have remained virtually unchanged for many decades while our world is changing rapidly, we are seeing a growing number of disengaged students. This disengagement increases the chances of students dropping out of school. For instance, high dropout rates, especially among Hispanic, black, and Native American students, were described as "the silent epidemic" in a recent research report for the Bill and Melinda Gates Foundation (Bridgeland, DiIulio, and Morison 2006). According to this report, nearly one-third of all public high school students drop out, and the rate is higher for minority students. In the report, when 467 high school dropouts were asked why they left school, 47 percent of them simply responded, "The classes were not interesting." We need to find ways (e.g., well-designed digital games and other immersive environments) to get our kids engaged, support their learning, and allow them to contribute fruitfully to society.

A second reason for using games as assessments is a pressing need for dynamic, ongoing measures of learning processes and outcomes. An interest in alternative forms of assessment is driven by dissatisfaction with and the limitations of multiple-choice items. In the 1990s, an interest in alternative forms of assessment increased with the popularization of what became known as authentic assessment. A number of researchers found that multiple-choice and other fixed-response formats substantially narrowed school curricula by emphasizing basic content knowledge and skills within subjects, and not assessing higher-order thinking skills (see, e.g., Kellaghan and Madaus 1991; Shepard 1991). As George Madaus and Laura O'Dwyer (1999) argued, though, incorporating performance assessments into testing programs is hard because they are less efficient, more difficult and disruptive to administer, and more time consuming than multiple-choice testing programs. Consequently, multiple choice has remained the dominant format in most K–12 assessments in our country. New performance assessments are needed that are valid, reliable, and automated in terms of scoring.

A third reason for using games as assessment vehicles is that many of them typically require a player to apply various competencies (e.g., creativity, problem solving, persistence, and collaboration) to succeed in the game. The competencies required to succeed in many games also happen to be the same ones that companies are looking for in today's highly competitive economy (Gee, Hull, and Lankshear 1996). Moreover, games are a significant and ubiquitous part of young people's lives. The Pew Internet and American Life Project, for instance, surveyed 1,102 youths between the ages of twelve and seventeen. They reported that 97 percent of youths—both boys (99 percent) and girls (94 percent)—play some type of digital game (Lenhart et al. 2008). Additionally, Mizuko Ito and her colleagues (2010) found that playing digital games with friends and family is a large as well as normal part of the daily lives of youths. They further observed that playing digital games is not solely for entertainment purposes. In fact, many youths participate in online discussion forums to share their knowledge and skills about a game with other players, or seek help on challenges when needed.

In addition to the arguments for using games as assessment devices, there is growing evidence of games supporting learning (see, e.g., Tobias and Fletcher 2011; Wilson et al. 2009). Yet we need to understand more precisely how as well as what kinds of knowledge and skills are being acquired. Understanding the relationships between games and learning is complicated by the fact that we don't want to disrupt players' engagement levels during gameplay. As a result, learning in games has historically been assessed indirectly and/or in a post hoc manner (Shute and Ke 2012; Tobias et al. 2011). What's needed instead is real-time assessment and support of learning based on the dynamic needs of players. We need to be able to experimentally ascertain the degree to which games can support learning, and how and why they achieve this objective.

This book presents the theoretical foundations of and research methodologies for designing, developing, and evaluating stealth assessments in digital games. Generally, stealth assessments are embedded deeply within games to unobtrusively, accurately, and dynamically measure how players are progressing relative to targeted competencies (Shute 2011; Shute, Ventura, et al. 2009). Embedding assessments within games provides a way to monitor a player's current level on valued competencies, and then use that information as the basis for support, such as adjusting the difficulty level of challenges or providing timely feedback. The term and technologies of stealth assessment are not intended to convey any type of deception but rather to reflect the invisible capture of gameplay data, and the subsequent formative use of the information to *help* learners (and ideally, help learners to help themselves).

There are four main sections in this report. First, we discuss problems with existing traditional assessments. We then review evidence relating to digital games and learning. Third, we define and then illustrate our stealth assessment approach with a set of assessments that we are currently developing and embedding in a digital game (*Newton's Playground*). The stealth assessments are intended to measure the levels of creativity, persistence, and conceptual understanding of Newtonian physics during game-play. Finally, we discuss future research and issues related to stealth assessment in education.

Problems with Current Assessments

Our country's current approach to assessing students (K–16) has a lot of room for improvement at the classroom and high-stakes levels. This is especially true in terms of the lack of support that standardized, summative assessments provide for students learning new knowledge, skills, and dispositions that are important to succeed in today's complex world. The current means of assessing students infrequently (e.g., at the end of a unit or school year for grading and promotion purposes) can cause various unintended consequences, such as increasing the dropout rate given the out-of-context and often irrelevant test-preparation teaching contexts that the current assessment system frequently promotes.

The goal of an ideal assessment policy/process should be to provide valid, reliable, and actionable information about students' learning and growth that allows stakeholders (e.g., students, teachers, administrators, and parents) to utilize the information in meaningful ways. Before describing particular problems associated with current assessment practices, we first offer a brief overview of assessment.

Assessment Writ Large

People often confound the concepts of *measurement* and *assessment*. Whenever you need to measure something accurately, you use an appropriate tool to determine how tall, short, hot, cold, fast, or slow something is. We measure to obtain information (data), which may or may not be useful, depending on the accuracy of the tools we use as well as our skill at using them. Measuring things like a person's height, a room's temperature, or a car's speed is technically not an assessment but rather the collection of information relative to an established standard (Shute 2009).

Educational Measurement

Educational measurement refers to the application of a measuring tool (or standard scale) to determine the degree to which important knowledge, skills, and other attributes have been or are being acquired. It involves the collection and analysis of learner data. According to the National Council on Measurement in Education's Web site, this includes the theory, techniques, and instrumentation available for the measurement of educationally relevant human, institutional, and social characteristics. A test is education's equivalent of a ruler, thermometer, or radar gun. But a test does not typically improve learning any more than a thermometer cures a fever; both are simply tools. Moreover, as Catherine Snow and Jacqueline Jones (2001) point out, tests alone cannot enhance educational outcomes. Rather, tests can guide improvement (given that they are valid and reliable) if they motivate adjustments to the educational system (i.e., provide the basis for bolstering curricula, ensure support for struggling learners, guide professional development opportunities, and distribute limited resources fairly).

Again, we measure things in order to get information, which may be quantitative or qualitative. How we choose to *use* the data is a different matter. For instance, back in the early 1900s, students' abilities and intelligence were extensively measured. Yet this wasn't done to help them learn better or otherwise progress. Instead, the main purpose of testing was to track students into appropriate paths, with the understanding that their aptitudes were inherently fixed. A dominant belief during that period was that intelligence was part of a person's genetic makeup, and thus testing was aimed specifically at efficiently assigning students into high, middle, or low educational tracks according to their supposedly innate mental abilities (Terman 1916). In general, there was a fundamental shift to practical education going on in the country during the early 1900s, countering "wasted time" in schools while abandoning the classics as useless and inefficient for the masses (Shute 2007). Early educational researchers and administrators inserted the metaphor of the school as a "factory" into the national educational discourse (Kliebard 1987). The metaphor has persisted to this day.

Assessment

Assessment involves more than just measurement. In addition to systematically collecting and analyzing information (i.e., measurement), it also involves interpreting and acting on information about learners' understanding and/or performance relative to educational goals. Measurement can be viewed as a subset of assessment.

As mentioned earlier, assessment information can be used by a variety of stakeholders and for an array of purposes (e.g., to help improve learning outcomes, programs, and services as well as to establish accountability). There is also an assortment of procedures associated with the different purposes. For example, if your goal was to enhance an individual's learning, and you wanted to determine that individual's progress toward an educational goal, you could administer a quiz, view a portfolio of the student's work, ask the student (or peers) to evaluate progress, watch the person solve a complex task, review lab reports or journal entries, and so on.

In addition to having different purposes and procedures for obtaining information, assessments may be differentially referenced or interpreted–for instance, in relation to normative data or a criterion. Norm-referenced interpretation compares learner data to that of other individuals or a larger group, but can also involve comparisons to oneself (e.g., asking people how they are feeling and getting a "better than usual" response is a norm-reference interpretation). The purpose of norm-referenced interpretation is to establish what is typical or reasonable. On the other hand, criterion-referenced interpretation involves establishing what a person can or cannot do, or typically does or does not do—specifically in relation to a criterion. If the purpose of the assessment is to support personal learning, then criterion-referenced interpretation is required (for more, see Nitko 1980).

This overview of assessment is intended to provide a foundation for the next section, where we examine specific problems surrounding current assessment practices.

Traditional Classroom Assessments Are Detached Events

Current approaches to assessment are usually divorced from learning. That is, the typical educational cycle is: teach; stop; administer test; go loop (with new content). But consider the following metaphor representing an important shift that occurred in the world of retail outlets (from small businesses to supermarkets to department stores), suggested by James Pellegrino, Naomi Chudhowsky, and Robert Glaser (2001, 284). No longer do these businesses have to close down once or twice a year to take inventory of their stock. Rather, with the advent of automated checkout and bar codes for all items, these businesses have access to a continuous stream of information that can be used to monitor inventory and the flow of items. Not only can a business continue without interruption; the information obtained is also far richer than before, enabling stores to monitor trends and aggregate the data into various kinds of summaries as well as to support real-time, just-in-time inventory management. Similarly, with new assessment technologies, schools should no longer have to interrupt the normal instructional process at various times during the year to administer external tests to students. Assessment instead should be continual and invisible to students, supporting real-time, just-in-time instruction (for more, see Shute, Levy, et al. 2009).

Traditional Classroom Assessments Rarely Influence Learning

Many of today's classroom assessments don't support deep learning or the acquisition of complex competencies. Current classroom assessments (referred to as "assessments of learning") are typically designed to judge a student (or group of students) at a single point in time, without providing diagnostic support to students or diagnostic information to teachers. Alternatively, assessments (particularly "assessments for learning") can be used to: support the learning process for students and teachers; interpret information about understanding and/or performance regarding educational goals (local to the curriculum, and broader to the state or common core standards); provide formative compared to summative information (e.g., give useful feedback during the learning process rather than a single judgment at the end); and be responsive to what's known about how people learn—generally and developmentally.

To illustrate how a classroom assessment may be used to support learning, Valerie Shute, Eric Hansen, and Russell Almond (2008) conducted a study to evaluate the efficacy of an assessment for learning system named ACED (for "adaptive content with evidence-based diagnosis"). They used an evidence-centered design approach (Mislevy, Steinberg, and Almond 2003) to create an adaptive, diagnostic assessment system that also included instructional support in the form of elaborated feedback. The key issue examined was whether the inclusion of the feedback into the system impairs the quality of the assessment (relative to validity, reliability, and efficiency), and does in fact enhance student learning. Results from a controlled evaluation testing 268 high-school students showed that the quality of the assessment was unimpaired by the provision of feedback. Moreover, students using the ACED system showed significantly greater learning of the content (geometric sequences) compared with a control group (i.e., students using the system but without elaborated feedback—just correct/incorrect feedback). These findings suggest that assessments in other settings (e.g., state-mandated tests) can be augmented to support student learning with instructional feedback without jeopardizing the primary purpose of the assessment.

Traditional Assessment and Validity Issues

Assessments are typically evaluated under two broad categories: reliability and validity. Reliability is the most basic requirement for an assessment and is concerned with the degree to which a test can consistently measure some attribute over similar conditions. In assessment, reliability is seen, for example, when a person scores really high on an algebra test at one point in time and then scores similarly on a comparable test the next day. In order to achieve high reliability, assessment tasks are simplified to independent pieces of evidence that can be modeled by existing measurement models.

An interesting issue is how far this simplification process can go without negatively influencing the validity of the test. That is, in order to remove any possible source of construct-irrelevant variance and dependencies, tasks can end up looking like decontextualized, discrete pieces of evidence. In the process of achieving high reliability, which is important for supporting high-stakes decision making, other aspects of the test may be sacrificed (e.g., engagement and some types of validity).

Another aspect that traditional, standardized assessments emphasize is dealing with operational constraints (e.g., the need for gathering and scoring sufficient pieces of evidence within a limited administration time and budget). In fact, many of the simplifications described above could be explained by this issue along with the current state of certain measurement models that do not easily handle complex interactions among tasks, the presence of feedback, and student learning during the test.

Validity, broadly, refers to the extent to which an assessment actually measures what it is intended to measure. Here are the specific validity issues related to traditional assessment.

Face Validity
Face validity states that an assessment should intuitively "appear" to measure what it is intended to measure. For example,

reading some excerpted paragraphs on an uninteresting topic and answering multiple-choice questions about it may not be the best measure for reading comprehension (i.e., it lacks good face validity). As suggested earlier, students need to be assessed in meaningful environments rather than filling in bubbles on a prepared form in response to decontextualized questions. Digital games can provide such meaningful environments by supplying students with scenarios that require the application of various competencies, such as reading comprehension and problem-solving skill.

Predictive Validity

Predictive validity refers to an assessment predicting future behavior. Today's large-scale, standardized assessments are generally lacking in this area. For example, a recent report from the College Board found that the SAT only marginally predicted college success beyond high school GPA at around $r = 0.10$ (Korbin et al. 2008). This means that the SAT scores contribute around 1 percent of the unique prediction of college success after controlling for GPA information. Other research studies have shown greater incremental validity of noncognitive variables (e.g., pyschosocial) over SAT and traditional academic indicators like GPA in predicting college success (see, e.g., Robbins et al. 2004).

Consequential Validity

Consequential validity refers to the effects of a particular assessment on societal and policy decisions. One negative side effect of the No Child Left Behind (NCLB 2002) initiative, with its heavy focus on accountability, has been teachers "teaching to the test." That is, when teachers instruct content that is relevant to answering items on a test but not particularly relevant for solving real-world problems, this reduces student engagement in school, and in turn, that can lead to increased dropout rates (Bridgeland, DiIulio, and Morison 2006). Moreover, the low predictive validity of current assessments can lead to students not getting into college due to low scores. But the SAT and similar test scores are still being used as the main basis for college admission decisions, which can potentially lead to some students missing opportunities at fulfilling careers and lives, particularly disadvantaged youths.

To illustrate the contrast between traditional and new performance-based assessments, consider the attribute of conscientiousness. Conscientiousness can be broadly defined as the motivation to work hard despite challenging conditions—a disposition that has consistently been found to predict academic achievement from preschool to high school to the postsecondary level and adulthood (see, e.g., Noftle and Robins 2007; O'Connor and Paunonen 2007; Roberts et al. 2004). Conscientiousness measures, like most dispositional measures, are primarily self-report (e.g., "I work hard no matter how difficult the task"; "I accomplish my work on time")—a method of assessment that is riddled with problems. First, self-report measures are subject to "social desirability effects" that can lead to false reports about behavior, attitudes, and beliefs (see Paulhaus 1991). Second, test takers may interpret specific self-report items differently (e.g., what it means "to work hard"), leading to unreliability and lower validity (Lanyon and Goodstein 1997). Third, self-report items often require that individuals have explicit knowledge of their dispositions (see, e.g., Schmitt 1994), which is not always the case.

Good games, coupled with evidence-based assessment, show promise as a vehicle to dynamically measure conscientiousness and other important competencies more accurately than traditional approaches (see, e.g., Shute, Masduki, and Donmez 2010). These evidence-based assessments can record and score multiple behaviors as well as measurable artifacts in the game that pertain to particular competencies. For example, various actions that a player takes within a well-designed game can inform conscientiousness—how long a person spends on a difficult problem (where longer equals more persistent), the number of failures and retries before success, returning to a hard problem after skipping it, and so on. Each instance of these "conscientiousness indicators" would update the student model of this variable—and thus would be up to date and available to view at any time. Additionally, we posit that good games can provide a gameplay environment that can potentially *improve* conscientiousness, because many problems require players to persevere despite failure and frustration. That is, many good games can be quite difficult, and pushing one's limits is an excellent way to improve persistence, especially when accompanied by the great sense of satisfaction one gets on successful completion of a thorny problem (see, e.g., Eisenberg 1992; Eisenberg and Leonard 1980). Some students, however, may not feel engaged or comfortable with games, or cannot access them. Alternative approaches should be available for these students.

As can be seen, traditional tests may not fully satisfy various validity and learning requirements. In the next section we describe how digital games can be effectively used in education—as assessment vehicles and to support learning.

Digital Games, Assessment, and Learning

Digital games are popular. For instance, revenues for the digital game industry reached US $7.2 billion in 2007 (Fullerton 2008), and overall, 72 percent of the population in the United States plays digital games (Entertainment Software Association 2011). The amount of time spent playing games also continues to increase (Escobar-Chaves and Anderson 2008). Besides being a popular activity, playing digital games has been shown to be positively related to a variety of cognitive skills (on visual-spatial abilities, e.g., see Green and Bavelier 2007; on attention, e.g., see Shaw, Grayson, and Lewis 2005), openness to experience (Chory and Goodboy 2011; Ventura, Shute, and Kim 2012; Witt, Massman, and Jackson 2011), persistence (i.e., a facet of conscientiousness; Ventura, Shute, and Zhao, forthcoming), academic performance (e.g., Skoric, Teo, and Neo 2009; Ventura, Shute, and Kim 2012), and civic engagement (Ferguson and Garza 2011). Digital games can also motivate students to learn valuable academic content and skills, within and outside the game (e.g., Barab, Dodge, et al. 2010; Coller and Scott 2009; DeRouin-Jessen 2008). Finally, studies have shown that playing digital games can promote prosocial and civic behavior (e.g., Ferguson and Garza 2011).

As mentioned earlier, learning in games has historically been assessed indirectly and/or in a post hoc manner (see Shute and Ke 2012). What is required instead is real-time assessment and support of learning based on the dynamic needs of players. Research examining digital games and learning is usually conducted using pretest-game-posttest designs, where the pre- and posttests typically measure content knowledge. Such traditional assessments don't capture and analyze the dynamic, complex performances that inform twenty-first-century competencies. How can we both measure *and* enhance learning in real time? Performance-based assessments with automated scoring are needed. The main assumptions underlying this new approach are that: learning by doing (required in gameplay) improves learning processes and outcomes; different types of learning and learner attributes may be verified as well as measured during gameplay; strengths and weaknesses of the learner may be capitalized on and bolstered, respectively, to improve learning; and ongoing feedback can be used to further support student learning.

Evidence of Learning from Games

Below are three examples of learning from educational games. Preliminary evidence suggests that students can learn deeply from such games and acquire important twenty-first-century competencies.

Programming Skills in *NIU-Torcs*
The game *NIU-Torcs* (Coller and Scott 2009) requires players to create control algorithms to make virtual cars execute nimble

maneuvers and stay balanced. At the beginning of the game, players receive their own cars, which sit motionless on a track. Each student must write a C++ program that controls the steering wheel, gearshift, accelerator, and brake pedals to get the car to move (and stop). The program also needs to include specific maneuverability parameters (e.g., gas pedal, transmission, and steering wheel). Running their C++ programs permits students to simulate the car's performance (e.g., distance from the center line of the track and wheel rotation rates), and thus students are able to see the results of their programming efforts by driving the car in a 3-D environment.

NIU-Torcs was evaluated using mechanical engineering students in several undergraduate classrooms. Findings showed that students in the classroom using *NIU-Torcs* as the instructional approach (n = 38) scored significantly higher than students in four control group classrooms (n = 48) on a concept map assessment. The concept map assessment included questions spanning four progressively higher levels of understanding: the number of concepts recalled (i.e., low-level knowledge),

Figure 1
Screen capture of *NIU-Torcs*

the number of techniques per topic recalled, the depth of the hierarchy per major topic (i.e., defining features and their connections), and finally, connections among branches in the hierarchy (i.e., showing a deep level of understanding). Students in the *NIU-Torcs* classroom significantly improved in terms of the depth of hierarchy and connections among branches (i.e., deeper levels of knowledge) relative to the control group. Figure 1 shows a couple of screen shots from the *NUI-Torcs* game.

Understanding Cancer Cells with *Re-Mission*

Re-Mission (Kato et al. 2008) is the name of a video game in which players control a nanobot (named Roxxi) in a 3-D environment representing the inside of the bodies of young patients with cancer. The gameplay was designed to address behavioral issues that were identified in the literature and were seen as critical for optimal patient participation in cancer treatment. The video gameplay includes destroying cancer cells and managing common treatment-related adverse effects, such as bacterial infections, nausea, and constipation. Neither Roxxi nor any of the virtual patients die in the game. That is, if players fail at any point in the game, then the nanobot powers down and players are given the opportunity to retry the mission. Players need to complete missions successfully before moving on to the next level.

A study was conducted to evaluate *Re-Mission* at thirty-four medical centers in the United States, Canada, and Australia. A total of 375 cancer patients, thirteen to twenty-nine years old, were randomly assigned to the intervention (n = 197) or control group (n = 178). The intervention group played *Re-Mission* while the control group played *Indiana Jones and the Emperor's Tomb* (i.e., both the gameplay and interface were similar to *Re-Mission*). After taking a pretest, all participants received a computer either with *Indiana Jones and the Emperor's Tomb* (control group) or the same control group game plus the *Re-Mission* game (intervention group). The participants were asked to play the game(s) for at least one hour per week during the three-month study, and outcome assessments were collected at one and three months after the pretest. Game use was recorded electronically. Outcome measures included adherence to taking prescribed medications, self-efficacy, cancer-related knowledge, control, stress, and quality of life. Adherence, self-efficacy, and cancer-related knowledge were all significantly greater in the intervention group

Figure 2
Screen capture of *Re-Mission* game

compared to the control group. The intervention did not affect self-reported measures of stress, control, or quality of life. Figure 2 shows an opening screen of *Re-Mission*.

Taiga Park and Science Content Learning

Our last example illustrates how kids learn science content and inquiry skills within an online game called *Quest Atlantis: Taiga Park*. *Taiga Park* is an immersive digital game developed by Sasha Barab and his colleagues at Indiana University (Barab et al. 2007; Barab, Gresalfi, and Ingram-Goble 2010). *Taiga Park is* a beautiful national park where many groups coexist, such as the fly-fishing company, the Mulu farmers, the lumber company, and park visitors. In this game, Ranger Bartle calls on the player to investigate why the fish are dying in the Taiga River. To solve this problem, players are engaged in scientific inquiry activities. They interview virtual characters to gather information, and collect water samples at several locations along the river to measure water quality. Based on the collected information, players make a hypothesis and suggest a solution to the park ranger.

To move successfully through the game, players need to understand how certain science concepts are related to each other (e.g., sediment in the water from the loggers' activities causes an increase to the water temperature, which decreases the amount of dissolved oxygen in the water, which causes the fish to die). Also, players need to think systemically about how different social, ecological, and economic interests are intertwined in this park. In a controlled experiment, Barab and his colleagues (2010) found that middle-school students learning with *Taiga Park* scored significantly higher on the posttest (i.e., assessing knowledge of core concepts such as erosion and eutrophication) compared to the classroom condition ($p < 0.01$). The *Taiga Park*

Figure 3
Screen capture of *Taiga Park*

group also scored significantly higher than the control condition on a *delayed* posttest, thus demonstrating retention of the content relating to water quality ($p < 0.001$) in a novel task (thus better retention and transfer). The same teacher taught both treatment and control conditions. For a screen capture from *Taiga Park*, see figure 3.

As these examples show, digital games appear to support learning. But how can we more accurately measure learning, especially as it happens (rather than after the fact), and beyond content knowledge?

Assessment in Games

In a typical digital game, as players interact with the environment, the values of different game-specific variables change. For instance, getting injured in a battle reduces a player's health, and finding a treasure or another object increases a player's inventory of goods. In addition, solving major problems in games permits players to gain rank or "level up." One could argue that these are all "assessments" in games—of health, personal goods, and rank. But now consider monitoring educationally relevant variables at different levels of granularity in games. In addition to checking health status, players could check their current levels of systems-thinking skill, creativity, and teamwork, where each of these competencies is further broken down into constituent knowledge and skill elements (e.g., teamwork may be broken down into cooperating, negotiating, and influencing/leadership skills). If the estimated values of those competencies got too low, the player would likely feel compelled to take action to boost them.

One main challenge for educators who want to employ or design games to support learning is making valid inferences—about what the student knows, believes, and can do—at any point in time, at various levels, and without disrupting the flow of the game (and hence engagement and learning). One way to increase the quality and utility of an assessment is to use evidence-centered design (ECD), which informs the design of valid assessments and yields real-time estimates of students' competency levels across a range of knowledge and skills (Mislevy, Steinberg, and Almond 2003).

ECD is a conceptual framework that can be used to develop assessment models, which in turn support the design of valid assessments. The goal is to help assessment designers coherently align the claims that they want to make about learners as well as the things that learners say or do in relation to the contexts and tasks of interest (e.g., Mislevy and Haertel 2006; Mislevy, Steinberg, and Almond 2003; for a simple overview, see ECD for

Dummies by Shute, Kim, and Razzouk 2010). There are three main theoretical models in the ECD framework: competency, evidence, and task models.

Competency Model

What collection of knowledge, skills, and other attributes should be assessed? Although ECD can work with simple one-dimensional competency models, its strength comes from treating competency as multidimensional. Variables in the competency model describe the set of knowledge and skills on which inferences are based (see Almond and Mislevy 1999). The term *student model* is used to denote an instantiated version of the competency model—like a profile or report card, only at a more refined grain size. Values in the student model express the assessor's current belief about the level on each variable within the competency model, for a particular student.

Evidence Model

What behaviors or performances should reveal those competencies? An evidence model expresses how the student's interactions with and responses to a given problem constitute evidence about competency model variables. The evidence model attempts to answer two questions: (a) What behaviors or performances reveal targeted competencies; and (b) What's the statistical connection between those behaviors and the variable(s) in the competency model?

Task Model

What tasks or problems should elicit those behaviors that comprise the evidence? The variables in a task model describe features of situations that will be used to elicit performance. A task model provides a framework for characterizing or constructing situations with which a student will interact to supply evidence about targeted aspects of competencies. The main purpose of tasks or problems is to elicit evidence (observable) about competencies (unobservable). The evidence model serves as the glue between the two.

There are two main reasons why we believe that the ECD framework fits well with the assessment of learning in digital games. First, in digital games, people learn in action (Gee 2003; Salen and Zimmerman 2005). That is, learning involves continuous interactions between the learner and game, so learning is inherently situated in context. The interpretation of knowledge and skills as the products of learning therefore cannot be isolated from the context, and neither should assessment. The ECD framework helps us to link what we want to assess and what learners do in complex contexts. Consequently, an assessment can be clearly tied to learners' actions within digital games, and can operate without interrupting what learners are doing or thinking (Shute 2011).

The second reason that ECD is believed to work well with digital games is because the ECD framework is based on the assumption that assessment is, at its core, an evidentiary argument. Its strength resides in the development of performance-based assessments where what is being assessed is latent or not apparent (Rupp et al. 2010). In many cases, it is not clear what people learn in digital games. In ECD, however, assessment begins by figuring out just what we want to assess (i.e., the claims we want to make about learners), and clarifying the intended goals, processes, and outcomes of learning.

Accurate information about the student can be used to support learning. That is, it can serve as the basis for delivering timely and targeted feedback as well as presenting a new task or quest that is right at the cusp of the student's skill level, in line with flow theory (e.g., Csikszentmihalyi 1990) and Lev Vygotsky's (1978) zone of proximal development.

As discussed so far, there are good reasons for using games as assessment vehicles to support learning. Yet Diego Zapata-Rivera and Malcolm Bauer (2011) discuss some of the challenges relating to the implementation of assessment in games, such as the following:

• *Introduction of construct irrelevant content and skills*   When designing interactive gaming activities, it is easy to introduce content and interactions that impose requirements on knowledge, skill, or other attributes (KSA) that are not part of the construct (i.e., the KSAs that we are not trying to measure). That is, authenticity added by the context of a game may also impose demands on irrelevant KSAs (Messick 1994). Designers need to explore the implications for the type of information that will be gathered and used as evidence of students' performance on the KSAs that are part of the construct.

• *Interaction issues*   The nature of interaction in games may be at odds with how people are expected to perform on an assessment task. Making sense of issues such as exploring behavior, pacing, and trying to game the system is challenging, and has a direct link to the quality of evidence that is collected about student behavior. The environment can lend itself to interactions that may not be logical or expected. Capturing the types of behaviors that will be used as evidence and limiting other types of behaviors (e.g., repeatedly exploring visual or sound effects) without making the game dull or repetitive is a challenging activity.

• *Demands on working memory*   Related to both the issues of construct-irrelevant variance (i.e., when the test contains excess variance that is irrelevant to the interpreted construct; Messick 1989) and interaction with the game is the issue of demands that gamelike assessments place on students' working memory. By designing assessments with higher levels of interactivity and engagement, it's easy to increase cognitive processing demands in a way that can reduce the quality of the measurement of the assessment.

• *Accessibility issues* Games that make use of rich, immersive graphic environments can impose great visual, motor, auditory, and other demands on the player to just be able to interact in the environment (e.g., sophisticated navigation controls). Moreover, creating environments that do not make use of some of these technological advances (e.g., a 3-D immersive environment) may negatively affect student engagement, especially for students who are used to interacting with these types of games. Parallel environments that do not impose the same visual, motor, and auditory demands without changing the construct need to be developed for particular groups of students (e.g., students with visual disabilities).

• *Tutorials and familiarization* Although the majority of students have played some sort of video game in their lives, students will need support to understand how to navigate and interact with the graphic environment. Lack of familiarity with navigation controls may negatively influence student performance and student motivation (e.g., Lim, Nonis, and Hedberg 2006). The use of tutorials and demos can support this familiarization process. The tutorial can also be used as an engagement element (see, e.g., Armstrong and Georgas 2006).

• *Type and amount of feedback* Feedback is a key component of instruction and learning. Research shows that interactive computer applications that provide immediate, task-level feedback to students can positively contribute to student learning (e.g., Hattie and Timperley 2007; Shute 2008; Shute, Hansen, and Almond 2008). Shute (2008) reviews research on formative feedback and identifies the characteristics of effective formative feedback (e.g., feedback should be nonevaluative, supportive, timely, specific, multidimensional, and credible). Immediate feedback that results from a direct manipulation of objects in the game can provide useful information to guide exploration or refine interaction strategies. The availability of ongoing feedback may influence motivation and the quality of the evidence produced by the system. Measurement models need to take into account the type of feedback that has been provided to students when interpreting the data gathered during their interaction with the assessment system.

• *Handling dependencies among actions* Dependencies among actions/events can be complex to model and interpret. Assumptions of conditional independence required by some measurement models may not hold in complex interactive scenarios. Designing scenarios carefully can help reduce the complexity of measurement models. Using data-mining techniques to support evidence identification can also help with this issue.

In addition to these challenges, in order to make scalable assessments in games, we need to take into account operational constraints and support the need for assessment information by different educational stakeholders, including students, teachers, parents, and administrators. Stealth assessment addresses many of these challenges. The next section describes stealth assessment and offers a sample application in the area of Newtonian physics.

Stealth Assessment

Given the goal of using well-designed games to support learning in school settings and elsewhere, we need to ensure that the assessments are valid, reliable, and also unobtrusive (to keep engagement intact). The output from the assessments, however, should be transparent. That is, players should be aware of how they are doing relative to important competencies at any point in time. One way to meet these requirements is to use "stealth assessment" (Shute 2011; Shute, Ventura, et al. 2009). Stealth assessment refers to ECD-based assessments that are woven directly and invisibly into the fabric of the gaming environment. During gameplay, students naturally produce rich sequences of actions while performing complex tasks, drawing on the very skills or competencies that we want to assess (e.g., scientific inquiry skills and creativity). Evidence needed to assess the skills is thus provided by the players' interactions with the game itself (i.e., the processes of play), which can be contrasted with a typically singular outcome of an activity—the norm in educational environments.

Making use of this stream of gameplay evidence to assess students' knowledge, skills, and understanding (as well as beliefs, feelings, and other learner states and traits) presents problems for traditional measurement models used in assessment. First, in traditional tests the answer to each question is seen as an independent data point. In contrast, the individual actions within a sequence of interactions in a game are often highly dependent on one another. What one does in a particular game at one point in time, for example, affects subsequent actions later on. Second, in traditional tests, questions are frequently designed to measure particular, individual pieces of knowledge or a skill. Answering the question correctly is evidence that one may know a certain fact: one question—one fact. But by analyzing a sequence of actions within a quest (where each response or action provides incremental evidence about the current mastery of a specific fact, concept, or skill), stealth assessments within game environments can infer what learners know and do not know at any point in time. Now because we typically want to assess a whole cluster of skills and abilities from evidence coming from learners' interactions within a game, methods for analyzing the sequence of behaviors to infer these abilities are not as obvious. As suggested above, evidence-based stealth assessments can address these problems.

Stealth Assessment in *Newton's Playground*

We have designed a number of stealth assessment *mock-ups* for measuring competencies within different games, such as systems-thinking skills in *Taiga Park* (Shute, Masduki, and Donmez 2010), creative problem solving in *Oblivion* (Shute, Ventura, et al. 2009), and causal reasoning in the *World of Goo* (Shute and Kim 2011). What needs to be done now is to actually build stealth assessments directly within a digital game, as part of gameplay. In a current research project, we are doing just that. Before describing the game and stealth assessments, we first discuss the research project.

Research Project

One year ago, we received funding from the Bill and Melinda Gates Foundation to design, develop, and evaluate three evidence-based assessments embedded in a digital game. The three focal competencies are creativity, conscientiousness, and conceptual physics understanding. The game we originally selected within which to embed our stealth assessments was *Crayon Physics Deluxe*, developed by Petri Purho. But given some conceptual issues with the game (e.g., mass was equivalent to size of an object not density, and interactions involving smooth and grassy surfaces did not differ in terms of friction), in addition to problems getting the source code, we decided to recode the game from scratch and call it *Newton's Playground*. The new game uses the same physics engine as *Crayon Physics Deluxe* (Box2D) and has identical core game mechanics with *Crayon Physics Deluxe* (e.g., drawing physical objects to create forces that come alive in a 2-D environment). Developing *Newton's Playground* has enabled us to fix some of the problems that were present in the original game (e.g., filling in an object in *Newton's Playground* comprises increasing its mass, so two equally sized objects with differing degrees of filling will have different masses). Additionally, developing *Newton's Playground* ourselves has allowed us to embed assessment mechanics seamlessly within the game mechanics. Data are being dynamically collected in *Newton's Playground* from middle-school players' interactions in the game. These observable performance data inform our three focal competencies. We provide details of how this works in the next section.

We are currently starting the second year of the two-year project, which involves creating and evaluating three stealth assessments. For gameplay, we have developed around ninety new problems (divided into difficulty levels of easy, medium, and hard) using the level editor in our new game. These problems were carefully developed to suit our experimental needs (e.g., creating certain aspects of a problem, such as difficulty level and physics principles needed in the solution). We have begun pilot testing the problems to determine if they're suitable for our population and methodological requirements (e.g., adequate variability). Pilot work is currently being conducted with a sample of about forty middle-school students in Tallahassee, Florida.

In our second year, we will conduct two studies to evaluate the *validity* of the stealth assessments, examine *learning* from the game, and test the *scalability* of the stealth assessments to other games. Specifically, in the first study ($n = 150$), we will evaluate the validity of our three stealth assessments in *Newton's Playground*. Students will complete a pretest battery of traditional tests on our three focal competencies, interact with a pool of *Newton's Playground* problems (ninety carefully designed problems) over two two-hour sessions in the computer lab at their school, and complete a posttest on conceptual physics understanding. Students' competency levels will be estimated from their gameplay in *Newton's Playground*, and the competency estimates will be correlated with scores from the traditional tests (for examples of our external measures, see appendix 2). The results of the study will inform us as to the validity of the stealth assessments for the three focal competencies (creativity, conscientiousness, and conceptual physics) and provide preliminary evidence for conceptual physics learning in *Newton's Playground*.

In our second study during year two, we will employ one of our stealth assessment models (i.e., persistence, which is a main facet of conscientiousness) in a different learning environment. Students will interact with the second game over two one-hour sessions in the computer lab at their school. At the end of the sessions, the competency estimates for persistence will be compared to traditional tests to evaluate the validity of the assessments as well as the scalability of the persistence models (i.e., developed for one game and reused within another environment). This part of the project is important because developing the competency and evidence models for each of our three competencies is time consuming, and thus expensive. That is, the process of model creation involves more than six months of extensive literature review per competency, structuring all of the abstracted variables into graphic and statistical models, and then having experts in the respective areas evaluate the models for face validity. Being able to recycle the models would allow us to scale the stealth technologies.

We now turn our attention to the game we are using as the vehicle for our stealth assessment project.

Newton's Playground

Newton's Playground is a computer game that emphasizes 2-D physics simulations, including gravity, mass, kinetic energy, and transfer of momentum. The objective of each problem in *Newton's Playground* is to guide a green ball from a predetermined starting point to a red balloon (or balloons), which pops on contact and gives the student a trophy for the successfully completed level (and multiple trophies for multiple solutions). Everything obeys the basic rules of physics relating to gravity and Sir Isaac Newton's three laws of motion. The player can nudge the ball to the left and right (if the surface is flat), but the primary way to move the ball is by drawing physical objects on the screen that "come to life" once the object is drawn. For example, in the "golf problem" (see figure 4), the player must draw a golf club on a pin (i.e., little circle on the cloud) to make it swing down to hit the ball. In the depicted solution, the player also drew a ramp to prevent the ball from falling down a pit.

The speed of (and importantly, the impulse delivered by) the swinging golf club is dependent on the size/mass distribution of the club and the angle from which it was dropped to swing.

Figure 4
Golf problem in *Newton's Playground*

The ball will then fly at a certain speed, length, and trajectory. If drawn properly, the ball will hit the balloon.

The various problems in *Newton's Playground* require the player to create and use ramps, pendulums, levers, and so forth to move the ball. All solutions are drawn with colored markers using the mouse. In a number of cases the ball must go over a pit. If the ball falls into the pit, the player must start the problem over.

Players can replay a problem as often as they like—even after successfully solving it. Players get a silver trophy for a solution. If players solve a problem with just a one to two objects they receive a gold trophy. One motivation to replay a problem is to find even more elegant and creative solutions than were generated before. It is not uncommon for a player to revisit/replay particularly challenging problems multiple times, always striving for a better, more elegant solution.

Agents of Force and Motion

As noted before, *Newton's Playground* requires players to create and use the following devices (or what we have been calling "agents of force and motion") to help the ball reach the balloon:

1. *Ramp* A ramp can be employed to change the direction of the motion of the ball (or another object). In some cases, other tools (like a pendulum or nudge) are needed to get the ball to start moving.

2. *Lever* A seesaw or lever involves net torque. A lever rotates around a fixed point, usually called a fulcrum or pivot point. An object residing on a lever gains potential energy as it is raised.

3. *Pendulum* A swinging pendulum directs an impulse tangent to its direction of motion. The idealized pendulum is a specialized case of the physical pendulum for which the mass

distribution helps determine the frequency. One can draw a physical pendulum in *Newton's Playground*, and the motion will be determined by the mass distribution.

4. *Springboard* A springboard (or diving board) stores elastic potential energy provided by a falling weight. Elastic potential energy becomes kinetic as the weight is released.

5. *Pin* A pin allows the position of one body to be fixed in space. Like a nail, it supplies a force large enough to resist the motion of the point it is attached to. Two pins hold a body fairly immobile against a background; more pins increase the immobilization.

6. *Rope* Ropes generally transmit tension between objects and can act like trampolines, generating forces on objects by stretching the rope and then removing the force (by deleting objects) to produce upward momentum on the ball.

7. *Nudge* Left or right clicking on the ball in *Newton's Playground* allows the user to poke/nudge the ball into motion.

The next section introduces the three competency models and their associated indicators (i.e., evidence) in the *Newton's Playground* project. For each of the three competency models, we review the relevant literature and then present a coherent graphic model of the variables. In the graphic models, unobservable/theoretical variables are on the left and observable/measurable indicators are on the right (i.e., what a person does in the environment to inform the latent variables).

Conscientiousness Review and Competency Model

Over the past twenty years or so, conscientiousness has emerged as one of the most important competencies in academia (e.g.,

Poropat 2009) as well as the workforce (e.g., Roberts et al. 2007; Schmidt and Hunter 1998). Conscientiousness is a multifaceted competency that commonly includes tendencies related to being attentive, hardworking, careful, detail minded, reliable, organized, productive, and persistent (Noftle and Robins 2007; Roberts et al. 2005). It is also important to note that conscientiousness is not highly related to mathematics skill or verbal reasoning (Trapmann et al. 2007)—measures typically used to assess general cognitive ability. Thus conscientiousness is considered to be noncognitive (i.e., a person's level of conscientiousness is relatively independent from cognitive measures such as standardized tests like the ACT or SAT).

The independence of conscientiousness from intelligence means it can affect students with high or low levels of cognitive ability. For example, a person who has high cognitive ability but low conscientiousness may end up performing about the same in school as a person who is low on cognitive ability and high on conscientiousness. Conscientiousness therefore can be seen as an independent attribute that can help or hinder performance in school. It is unclear why certain people have higher or lower levels of conscientiousness. Conscientiousness does not appear to be related to socioeconomic status (Roberts et al. 2007), but has been shown to increase over one's lifetime (Roberts, Walton, and Viechtbauer 2006). In the next section we review the empirical evidence regarding the validity of conscientiousness.

Validity of Conscientiousness
A number of studies and meta-analyses have shown the significance of self-report measures of conscientiousness in predicting a variety of crucial outcomes while controlling for cognitive ability. Conscientiousness has consistently been found to predict

academic achievement from preschool (Abe 2005) to high school (Noftle and Robins 2007; Poropat 2009) to the postsecondary level (O'Conner and Paunonen 2007; Trapmann et al. 2007) and adulthood (e.g., De Fruyt and Mervielde 1996; Shiner, Masten, and Roberts 2003). Meta-analyses have found conscientiousness to be correlated with grades between $r = 0.21$ and 0.27, and as mentioned, the relationship is independent of intelligence (e.g., Noftle and Robins 2007; Poropat 2009; Robbins et al. 2004). Existing research suggests that the organizational aspects of conscientiousness (e.g., organizing and planning) show the weakest relationships with school achievement while the aspects representing goal completion, persistence, and productivity show the strongest relationships (e.g., Roberts et al. 2005). In the next section we describe research on the structural facets of conscientiousness.

Structural Models of Conscientiousness

Carolyn MacCann, Angela Duckworth, and Richard Roberts (2009) reviewed studies that examined the underlying structure of conscientiousness (Peabody and De Raad 2002; Perugini and Gallucci 1997; Roberts et al. 2004; Roberts et al. 2005; Saucier and Ostendorf 1999). These studies are summarized in table 1.

The unit of analysis is important to distinguish since some studies conducted factor analysis on adjective ratings while others used factor analysis on self-report ratings. Three factors were common to all five studies (orderliness, industriousness, and responsibility/reliability), a control factor emerged in four of the five studies, and decisiveness and conventionality factors emerged in two studies.

Regarding the MacCann, Duckworth, and Roberts (2009) study, confirmatory factor analysis uncovered eight facets: industriousness, perfectionism, tidiness, proactivity, control,

Table 1
Underlying structure of conscientiousness (MacCann, Duckworth, and Roberts 2009)

	Perugini and Gallucci (1997)	Saucier and Ostendorf (1999)	Peabody and De Raad (2002)	Roberts et al. (2004)	Roberts et al. (2005)	MacCann, Duckworth, and Roberts 2009
Sample	Undergraduates (40%), employees (60%)	US undergraduates, German community volunteers	6 studies: 2 × Italian, Hungarian, Dutch, Polish, Czech	Mostly undergraduates (89%)	Community sample	High-school students
Unitanalyzed	Adjectives ratings	Adjectives ratings	Adjectives ratings	Adjectives ratings	Self-report ratings	Self-report ratings
Analysis	EFA	EFA	Conceptual judgments	EFA	EFA	EFA/CFA
Factor 1	Meticulousness	Orderliness	Orderliness	Orderliness	Order	Tidiness + task planning
Factor 2	Superficiality	Industriousness	Work	Industriousness	Industriousness	Industriousness + perfectionism
Factor 3	Reliability	Responsibility	Responsibility	Reliability	Responsibility	
Factor 4	Recklessness		Impulse control	Impulse control	Self-control	Control + cautiousness

Table 1
(continued)

	Perugini and Gallucci (1997)	Saucier and Ostendorf (1999)	Peabody and De Raad (2002)	Roberts et al. (2004)	Roberts et al. (2005)	MacCann, Duckworth, and Roberts 2009
Factor 5		Decisiveness consistency		Decisiveness		Proactivity
Factor 6				Convention-ality	Traditionalism	
Factor 7			Persistence			Perseverance
Factor 8				Punctuality		
Factor 9				Formalness		
Factor 10					Virtue	

cautiousness, task planning, and perseverance. All facets related meaningfully to broad conscientiousness, while perseverance also overlapped with neuroticism. The facets of industriousness and proactivity showed a higher prediction of student absences and attainment of academic honors compared with the other facets. Based on table 1, we developed a competency model that is displayed in figure 5.

As can be seen, we refined the conscientiousness model to include only four facets: *persistence, perfectionism, organization*, and *carefulness*. Our persistence facet combines the industriousness and perseverance facets since they both imply the notion of "working hard despite failure." We kept perfectionism as a facet, created the organization facet, and then broke it down into two main subfacets: resource management and time management. We created the carefulness facet, which can be broken down into caution (i.e., being careful not to make mistakes) and control (i.e., tendency to not act impulsively). Finally, the figure includes *Newton's Playground* indicators that can be linked to each of the facets. We illustrate next the task modeling that we have done for one facet of conscientiousness: persistence.

Task Modeling for Persistence

Assessing persistence is primarily based on seeing how long players spend or persist on problems that they do not readily solve. The challenge in this assessment design is that we are never really sure what problem a student may or may not be able to solve. To address this issue, we created difficulty rubrics for problems to systematically manipulate problem difficulty. This allows us to incrementally increase the difficulty of problems to ensure that students will eventually get to problems they will have trouble solving. Difficulty indexes include the following:

Unobservables/constructs   **Observables/indicators**

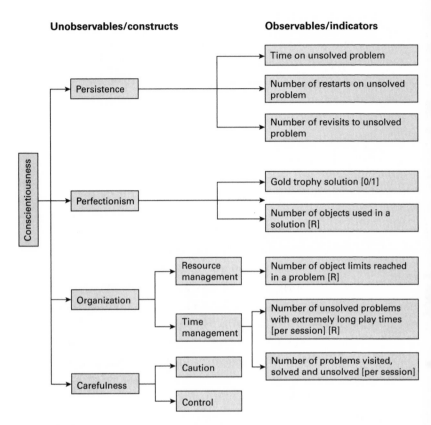

Figure 5
Competency model of conscientiousness with indicators from *Newton's Playground*

1. *Relative location of ball to balloon*   If the balloon is positioned *above* the ball in a problem, this makes the problem harder as it forces the player to use a lever, springboard, or pendulum to solve the problem (0–1 point).

2. *Obstacles*   This refers to the pathway between the ball and balloon. If the pathway is obstructed, this requires the player

to project the ball in a specific trajectory to obtain the balloon (0–2 points).

3. *Distinct agents of force/motion* (see previous section on agents of force and motion) A *Newton's Playground* problem may require just one or more agents to get the ball to the balloon (0–1 point).

4. *Novelty* This addresses whether a problem is novel relative to other problems played. For novel problems, the solution is not easily determined from experience with other problems (0–2 points).

Each problem is evaluated under all the rubrics to yield a total difficulty score (i.e., ranging from 0 to 6). Consider the "maze" problem in figure 6.

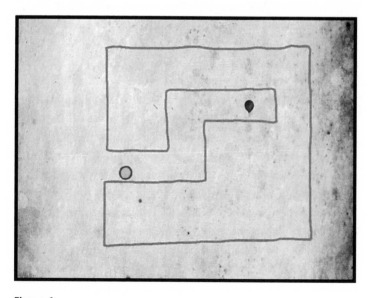

Figure 6
Maze problem in *Newton's Playground*

This problem gets a difficulty score of 5 as the balloon is above the ball (1 point), there's one obstacle—a narrow, crooked pathway—(1 point), at least two agents are typically needed to solve it (1 point), and there's no other problem like it in the game (2 points). Thus the maze problem would be a good task to assess persistence as it will likely be unsolvable by many students. We now turn our attention to creativity modeling.

Creativity Review and Competency Model

Creativity is generally defined as the ability to produce solutions or ideas that are both novel and effective (Lubart 1994). James Kaufman and Robert Sternberg (2007) have noted that most definitions of creativity (or creative problem solving) consist of three components: novelty, quality, and relevance. That is, creative solutions are novel, of high quality, and appropriate to the given task, or some variant of the task.

Various psychometric approaches exist to help understand and model creativity. According to Sternberg and Lubart (1992), there is a continuum between "less" contextualized approaches that focus on personal characteristics and "more" contextualized ones that include social-cultural variables that influence individuals' creativity. Typical less contextualized psychometric approaches explain creativity as a multifaceted construct that includes intelligence, knowledge, thinking styles, and personality traits. Robert McCrae (1987) stresses that the ability to think creatively in conjunction with an inclination to do so (i.e., disposition) leads to creative productions. Many other creativity researchers share similar views that creativity is a multifaceted construct that involves a convergence of multiple variables (e.g., Amabile 1983).

Validity of Creativity

Creativity has probably been one of the most elusive of all scientific constructs in psychology. For the past several decades, the field of creativity research has received criticism challenging its validity (Weisberg 1993). Nevertheless, many published reviews in creativity and innovation research show that interest in creativity is strong; it is also increasingly viewed as a key component relevant to academic success (e.g., Gronhaug and Kaufman 1988; Kaufman 2003; Runco 1997, 2002; Runco and Pritzker 1999; Sternberg 1988, 1999, 2006).

Divergent thinking tests are among the most popular techniques for measuring creativity in educational settings (Hunsaker and Callahan 1995; Runco 1992). These tests, also referred to as measures of ideational fluency, generally require students to provide as many responses as possible to prompts such as "List things that make noise" or "List things that have wheels." Among the most popular of the creativity tests are the Torrance Tests of Creative Thinking (Torrance 1974) and the Wallach and Kogan test (1965). The responses are usually scored for originality and fluency (number of responses) by expert raters. These tests have been shown to moderately predict important criteria such as school success (e.g., Okuda, Runco, and Berger 1991; Runco and Pritzker 1999; Sternberg 2006).

Structural Models of Creativity

Joy Paul Guilford (1956) conceptualized creativity as involving four facets of divergent thinking: *flexibility* (the ability to produce ideas from various categories or classes), *fluency* (the ability to rapidly produce a large number of ideas), *originality* (the ability to produce ideas that are unique, novel, and uncommon), and *elaboration* (the ability to develop the details of an idea and carry it out).

Flexibility has been recognized as an essential cognitive skill for creativity (Amabile 1983), and is defined as the ability to generate a varied pool of ideas by switching among categories and using remote associations (Nijstad et al. 2010; Runco 1986). Mark Runco (1986) discusses the significance of flexibility for creativity assessment: it distinguishes gifted from nongifted children better than fluency and originality, and it contributes to the predictive validity of divergent thinking tests with real-world criteria. People with a higher level of flexibility avoid using fixed problem-solving strategies, break perceptual sets, and make new connections among distant ideas. Even though the cognitive skills that are required for ideation (i.e., divergent thinking) are often considered as being synonymous with creativity, many caution that divergent thinking explains only *one* aspect of creativity, not the whole (e.g., Runco 2008). We agree with this position.

Openness to experience, one of the dimensions of the Big-Five factors, refers to a dispositional attribute that is characterized by an awareness of personal feelings and beliefs, receptivity to novel ideas, liberal values, intellectual curiosity, and fantasy (Berzonsky and Sullivan 1992). Therefore, individuals with higher degrees of openness to experience are described as imaginative, sensitive to aesthetics, curious, independent thinkers, and/or amenable to new ideas, experiences, and unconventional views (Costa and McCrae 1992). E. Paul Torrance (1974) explains that a creative individual tends to resist premature closure by keeping an open mind and considering a variety of information sources. A long line of research has supported the strong association of openness to experience with creativity or some aspects of creativity (Costa and McCrae 1992; Feist 1999; McCrae, 1987, 1996). For example, McCrae (1987) reported a significant association ($r = 0.4$) between divergent thinking and openness to experience.

A willingness to take risks (i.e., risk propensity) can be defined as the extent to which an individual takes an action knowing that there is uncertainty related to the potential payoff of the action (Dewett 2007). Risk taking is associated with an openness to change and new ideas (Madjar, Greenberg, and Chen 2011). A willingness to take risks (and knowing the possibility of failing) has been recognized as an essential trait of eminent scientists and artists throughout history. Sternberg and Lubart (1992) describe creative individuals as those who "buy low and sell high." They further argue that a willingness to take risks is a prerequisite for growth and creativity because one needs to go beyond what is commonly accepted as well as learn from various failings. Several studies have reported a positive association between a willingness to take risks and creativity (Glover 1977; Glover and Sautter 1977). For example, John Glover and Fred Sautter reported that a willingness to take risks was significantly correlated with flexibility and originality. A willingness to take risks has also been studied in the context of organizational innovations for many years (e.g., Dewett 2007; Kogan and Wallach 1964; MacCrimmon and Wehrung 1990). For example, Nora Madjar, Ellen Greenberg, and Zheng Chen (2011) found that a willingness to take risks is a significant contributor to individuals' radical creativity and innovation. Based on the literature, we have developed a competency model of creativity, displayed here in figure 7.

As can be seen, the model broadly splits creativity into cognitive and dispositional facets. The cognitive facet primarily refers to the ability to be creative in problem-solving tasks and is generally knowledge dependent. The dispositional facet refers

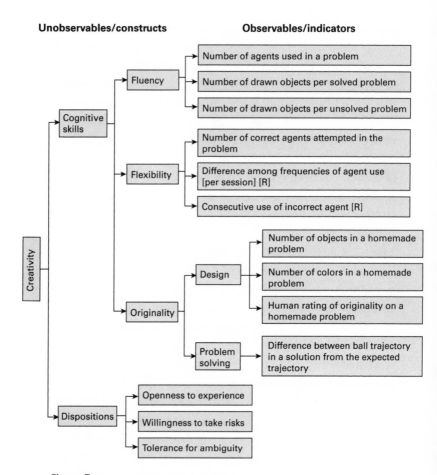

Figure 7
Competency model of creativity with indicators from *Newton's Playground*

to creative dispositions (e.g., openness) that are not necessarily related to cognitive ability or domain. Within the cognitive facet there are three subfacets: fluency, flexibility, and originality (note that we removed elaboration because there were no unique indicators for it). Based on the constraints of *Newton's Playground*, we will only focus on the cognitive facet of creativity. Among the cognitive facets, we now illustrate our task modeling on one of the variables: originality.

Task Modeling for Originality
Originality is assessed in terms of the uniqueness of a student's solution. This can be determined by seeing what agent(s) of force and motion a student used in a solution, or what trajectory the ball traveled in a solution. Consider the problem that we call *spider web* (see figure 8).

This problem can be solved with the lever, pendulum, springboard, and/or pulley. What we expect to see is that players will solve the problem (and others like it) in multiple and varied ways. This will be encouraged via the instructions we give to the students in the experiment. The instructions are:

You will have one hour to solve a pool of problems. Your goal is to solve as many of the problems, in as many awesome ways as you can. The tools we taught you will come in handy for many problems. Feel free, however, to solve any problem in whatever way you like. You also have the freedom to jump around and solve problems in any order that you like. For example, if you get stuck on a problem, you can leave it and come back to it later.

Each problem can be solved with one or more of your new tools. Each time you solve a problem, you'll be given a trophy. Your goal is to have as many trophies as possible. Again, if you get stuck on a problem, you can always leave that problem and go to another one. Just press the "escape" key and follow the directions to exit the problem. You can

Figure 8
Four agents in *Newton's Playground*

return to any problem as many times as you like. If you have any questions about how to play the game, please ask. Have fun!

When each session is complete, we will compare student ball trajectories in problems to expected trajectories to determine original solutions. Additionally, we will require students to create new levels after several hours of *Newton's Playground* gameplay. Students will create levels in the *Newton's Playground* level editor (i.e., the tool we use to make levels). We plan on hand scoring levels for originality as well as developing an algorithm for rating the originality of the level based on its features (e.g., number of different colors, number of objects, and number of obstacles between the ball and balloon). Next we discuss our model for conceptual physics understanding.

Conceptual Physics Review and Competency Model

Physics engines are becoming pervasive in gaming environments, providing a sense of realism in a game (e.g., Havok engine). Within these gaming environments, players can experiment with principles of physics such as impulse, inertia, vector addition, elastic collision, gravity, velocity, acceleration, free fall, mass, force, and projectile motion. The degree to which players apply these principles correctly in the game can be evidence of a conceptual understanding of physics. Specifically, players successfully drawing and applying the *Newton's Playground* agents of force and motion during problem solution will provide evidence related to knowing associated physics principles.

Over the past several decades it has become clear that many students who have achieved acceptable grades in one or more physics courses, actually have limited understanding of the physics involved (Halloun 1996; Swann 1950). Research in this area has shown that there are a number of routes to a passing grade that fail to develop an appreciation of physical principles and, more important, do not remove erroneous notions of how the world works from the students' understanding (e.g., Crouch and Mazur 2001; Hake 1998; Halloun and Hestenes 1985; McDermott 1993). This has led to widespread adoption of the text *Conceptual Physics* by Paul Hewitt (2009), currently in its eleventh edition, and the development of two instruments, the Force Concept Inventory (Hestenes, Wells, and Swackhamer 1992) and the Mechanics Baseline Test (Hestenes and Wells 1992), now widely used to compare student mastery of the concepts of mechanics between instructional approaches and courses. Recognition of the problem has also led to a renewed interest in the mechanisms by which physics students make the transition from naive or folk physics to Newtonian physics (diSessa 1982) as well as the possibility of video gameplaying assisting in the process (White 1994).

Based on foundational conceptual physics (e.g., Feynman 1964; Feynman, Leighton, and Sands 1964; Hewitt 2009), we interpret competency in conceptual physics to involve the following:

1. *Conceptual understanding of Newton's three laws of motion*   Newton's three laws of motion provide a conceptual understanding of how objects interact in the environment. The first law tells us that an object at rest stays at rest in the absence of any unbalanced forces, and an object in motion stays in motion in a straight line with unchanging speed in the absence of any forces. The second law ($F = ma$) tells us how the motion of the particle (object) evolves when it experiences a nonzero net force. Here F is the net force applied (i.e., the vector sum of all the forces acting on the object), m is the mass of the object, and a is the object's acceleration. The net force applied to an object thus produces a proportional acceleration. That is, if an object is accelerating, then there is a net nonzero force on it. In simple terms, it takes less force to produce the same acceleration of an object that has less mass compared to one with more mass. The third law states that for every action, there is an equal and opposite reaction. This can be illustrated by hitting a tree with a baseball bat. The force exerted on the tree by the swinging bat is equal to the force exerted back on the bat (and the person holding the bat) by the tree.

2. *Conceptual understanding of potential and kinetic energy*   Potential energy exists when a force acts on an object to restore the object to its resting point (or "lower energy configuration"). For example, when a springboard (like in *Newton's Playground*) is

bent downward, it exerts an upward force to return to its unbent position. The action of bending the springboard down requires energy, and the work done by the springboard in returning it to its resting point is considered stored as potential energy. When the bent springboard is released, the stored energy will be converted into kinetic energy.

3. *Conceptual understanding of conservation of angular momentum or torque* The angular momentum of a system of objects about any point of reference can be computed from the position and momentum of each of the objects. A useful example is a pendulum. When a pendulum with a long arm swings, it will accelerate faster than a pendulum with a shorter arm. Additionally, the mass of the pendulum will affect the force that a pendulum will exert (larger mass = more force). Finally, the position from which the pendulum is dropped (maximum = perpendicular or ninety degrees relative to floor) will affect the speed the pendulum moves when it is swung. Figure 9 shows the short version of our competency model for conceptual physics as it pertains to *Newton's Playground* (for the full model of physics principles, see appendix 1). As can be seen, the model includes Newton's three laws, potential and kinetic energy, and conservation of angular momentum or torque.

Newton's three laws is a parent principle in the model since it is pervasive in almost all problems in *Newton's Playground*. The successful use of each agent is an indicator of Newton's three laws. Additionally, there are indicators that inform each agent and principle.

Task Modeling for Conceptual Physics
As with the other competencies, all *Newton's Playground* problems require the player to use one or more agents of force and

Unobservables/constructs   **Observables/indicators**

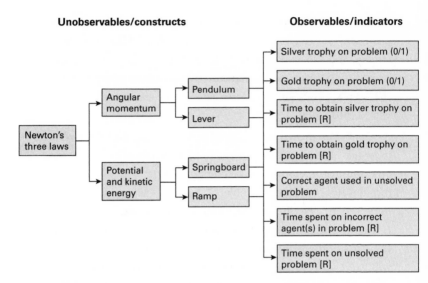

Figure 9
Competency model of conceptual physics with indicators from *Newton's Playground*

motion in the solution. Successful solutions therefore inform one or more of the competencies that we hope to develop in the student. As an illustration, consider the problem called *ballistic pendulum*, shown in figure 10.

This problem requires the student to create a pendulum shape with sufficient mass, and positioned so that the pendulum will fall down and "kick" the ball into a trajectory with sufficient force to slam into the red balloon (the figure on the right shows the ball en route to the balloon). Successfully solving this problem, in line with the competency model, suggests that the student has an intuitive grasp of the concepts of torque along with linear and angular momentum. It is necessary to correctly implement all the indicators to successfully get the ball to hit the target.

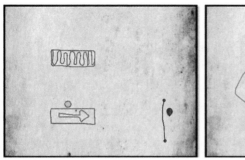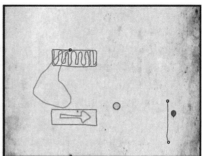

Figure 10
Ballistic pendulum problem (left) and solution (right)

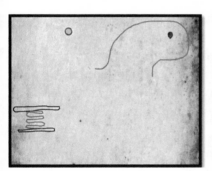

Figure 11
Diving board problem (left) and diving board solution (right)

Incidentally, the ballistic pendulum is an experiment often done in introductory physics courses in high school or college.

The springboard is a variant of the lever in which one flat board rests on another object that is pinned in place yet hangs over one edge. Figure 11 depicts the problem called diving board. When a weight is dropped from a height (or affixed) on to the free end of the springboard, the edge acts as an instantaneous

axis of rotation and the board experiences an angular acceleration. This can be used to launch objects up into space. It requires
a knowledge of potential and kinetic energy, and conservation
of angular momentum. Again, it is essential to correctly apply
the indicators to successfully get the object launched into the
tunnel to reach the balloon.

**Relation of Physics Indicators to Conscientiousness and Creativity
Indicators**

The indicators of conceptual physics understanding differ from
those of the other competencies in that they must be experi-
enced and learned. Also, they are domain specific. A measure—
such as the number of attempts to solve a problem—might
indicate a high level of persistence, but may also be consistent
with a lesser understanding of physics. In addition, really creative, single-object solutions in *Newton's Playground* may come
about through insight into physical principles, or more simply
by extensive trial and error.

The way we plan to resolve these issues is to model all relationships, within and among the three competency models, with
evidence models. We are using Bayesian networks to establish
the conditional probability relationships among the variables
within each competency model and some of the relationships,
as described above, between competency models.

Capturing Performance Data
So how do we capture the performance data that come from
gameplay, and use them to inform our three competency models? We have embedded code within the game that uses relevant
gameplay data to automatically identify agents and generate

evidence indicators. The complexity of this code varies depending on the indicator being generated. For example, it is simple to generate quantitative indicators such as the time spent on a problem (persistence) or number of solutions per problem (fluency). The aforementioned indicators can be generated using a *game-timer* variable and *counter* variable, respectively. Table 2 displays our current set of features used in our automated agent identification scheme.

These features require multiple queries to the physics engine used to model the game physics and a heuristic analysis of gameplay events. Consider, for instance, the pendulum agent. To test

Table 2

Features for identifying agents of force and motion in *Newton's Playground*

Agents	Features
Ramp	1. Number of bends (or tubes [i.e., tortuosity]) 2. Angle of each bend 3. Length of ramp
Lever	1. Length of the lever 2. Position of fulcrum 3. Height through which object falls before hitting lever 4. Mass of object 5. Location of the dropped object on lever (distance from fulcrum)
Pendulum	1. Angle of pendulum relative to horizontal fulcrum 2. Length between the axis point and the fulcrum (moment of inertia) 3. Mass (important when the pendulum hits something) 4. Position of pin
Springboard	1. Length of springboard 2. Mass of the object to weight it down 3. Position of the ball at release 4. Delete object or let fall off springboard 5. Angle of springboard at release

if a pendulum is striking the ball, our scheme queries the physics engine to determine if there are any objects contacting the ball, checks if any of these objects are attached by a single pin (i.e., allowing the object to freely rotate), and then determines if there is any change to the ball's trajectory resulting from the contact. If all three criteria are satisfied, our scheme generates a pendulum strike indicator.

To gauge the accuracy of our scheme during pilot testing, we are choosing a random selection of gameplay sessions and utilizing the "replay" feature of the game to perform a manual (visual) analysis of the indicators exhibited in each session. We are then comparing the indicators determined from the manual analysis with those automatically generated by our stealth assessment scheme. Modifications to the code for the automatically generated indicators will be made to align with human classifications, and eventually indicator classifications below 80 percent accuracy (relative to human evaluation) will not be implemented.

Example of an Evidence Model for Creativity

In general, the functional relationships among the competency and evidence models (i.e., indicators—obtained automatically via code in the game) can be presented as conditional probabilities by using a Bayesian network approach. To illustrate this, in our current model for creativity, the marginal probability of each level of the competency variables is initially set to roughly 33 percent, which is "uninformative" (see figure 12).

In some cases, like the number of drawn objects, we'd expect elegant (i.e., single object) solutions to be less frequently occurring as they are difficult to achieve (which influences our prior estimates). The difficulty and discrimination values of the indicators are also initially set to intermediate values because we

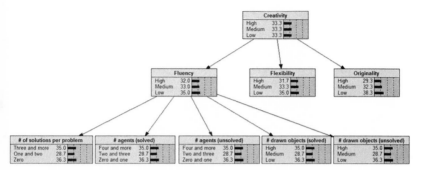

Figure 12
Competency model and evidence model for creativity—prior probabilities

do not yet have empirical data to know how those indicators actually function in our assessment, even though we have established some "difficulty indexes" based on problem characteristics. Once we collect students' data, the probability distributions will change, specific to our population.

To portray how the Bayesian network accumulates evidence and passes the information to the student model (i.e., the competency model that is specific to a student), we provide an example of a student's performance in the game (see figure 13).

The probabilities in the model will be refined based on responses from many students (i.e., many hours of gameplay data). Subsequently, instantiation of one student's evidence is used to infer values for latent variables. So after a two-hour session with the game, the student has generated, on average, three or more solutions per problem (see gray rectangle on lower left of figure). Yet the student only used, on average, two specific agents per solution attempt (i.e., lever and ramp—for successful and unsuccessful solutions). The student has also been judged to

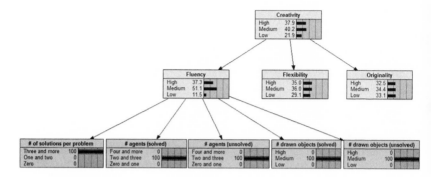

Figure 13
Competency model and evidence model for creativity—posterior probabilities

be at a "medium level" regarding the average number of objects created, for both solved and unsolved problems, relative to the population's performance data. All these fluency indicators (which are measurable) provide information about the number of solutions and objects that a student creates. In other words, the more solutions and objects, the more "fluent," in line with our definition of one of the main facets of creativity.

Once this evidence of the student's performance is incorporated into the student model, note that the level on the fluency node shifts toward the medium level (i.e., increasing from 33 to 51.1 percent). Moreover, the estimation about the student's overall creativity inches toward medium (i.e., increasing from 33 to 40.2 percent), while estimations for the other facets (i.e., flexibility and originality) do not change much. An operational version of this model also may include a variety of indicators not included here, such as indicators generated from pattern analysis processes from trace data across students and tasks, and indicators that supply evidence for more than one latent variable.

Newton's Playground Study Procedure

Each student will participate in the study for six hours (tested in groups of fifteen to twenty). This involves four hours of gameplay and two hours of testing (the latter occurring during the first and last sessions). During the first session, in addition to completing the set of external tests related to our three competencies, students will go through tutorials for the game. One of the challenges of conducting an assessment study with a game is that it requires the player to be comfortable with the mechanics of the game. In order to speed up this familiarization process, we have developed introductory videos as tutorials to teach students about various agents of force and motion. These tutorials dynamically illustrate how to draw each agent to solve a simple problem (during gameplay, students have the option to watch any agent-drawing video at any time). Once the students have been trained with the agents, they will play *Newton's Playground*.

Each gaming session will consist of an adequate pool of *Newton's Playground* problems that vary by difficulty levels (i.e., easy, medium, and hard). The difficulty level will be initially assigned based on our indexes described earlier, but may be adjusted based on performance data across the sample (e.g., a problem we assign as "easy" may only be solved by a small fraction of the students, thereby making it "difficult" in reality).

Some problems can be solved with multiple agents while others are intended to be solved by just one agent. For example, in the shark problem (see figure 14), the student can only use a lever to get the ball to the balloon (i.e., the problem was created explicitly to preclude other types of solutions and includes a built-in fulcrum via the shark's fin).

Figure 14
Shark problem and solution focusing on understanding lever mechanics

If a student cannot solve a particular problem, that student has the option to skip the problem and come back to it later. And as mentioned, the last gameplay session will contain a level-making activity where students will create a level relating to a principle of Newtonian physics. Students will be encouraged to create a level unlike any they have seen before to elicit creativity.

Discussion and Future Research

As we discussed at the beginning of this book, traditional assessments are often too simplified, abstract, and decontextualized to suit current and future education needs. We need new assessments that measure what students actually can *do* with the knowledge and skills obtained inside and outside of school. Digital games can provide meaningful assessment environments by providing students with problems that require the application of various competencies. We also presented an assessment methodology that enables us to develop tasks in digital games using the principles of ECD. These tasks are designed to elicit specific performance data, which are then statistically linked to our focal competencies.

The first and most important step of this research project will be the determination of the validity of our stealth assessments. We will also be examining any learning of conceptual physics that ensues from *Newton's Playground* gameplay. If, in fact, the stealth assessments accurately estimate the focal competencies relative to existing measures, and learning does occur after four hours of gameplay, then the next step is to examine scalability.

That is, what are the costs and benefits of recycling ECD-based models in different games to assess the same kinds of competencies? These issues are being studied in our research project described herein. If we find that our stealth assessment methodology yields valid and reliable information, and is scalable, we plan to make the process as well as models broadly available to the community so that the work will continue and grow.

The research can expand in a number of general directions. First, we (and/or others) can explore the development of stealth assessments for other competencies that have been shown to play crucial roles in academic (and life) success (e.g., communication skills, computational thinking, empathy, civic engagement, problem-solving skill, and teamwork). Second, we can look at the development of stealth assessments relating to content that is directly aligned with the common core standards (e.g., mathematics modeling, probability, or reading comprehension). Third, we can push the bounds of our stealth assessments relative to implementing the models in additional digital games as well as other digital learning environments to determine the range of environments that may employ the same competency and evidence models, for a scalable, cost-effective, and engaging solution to the assessment of complex competencies. And fourth, we can examine any added value of including exploratory, data-mining methods to stealth assessment's more theoretically driven approach relative to the quality of the assessment.

Regarding future research related to learning, stealth assessment has the potential to be quite useful for diagnostic purposes due to the fine-grained analysis of student behavior in situated contexts. In addition, real-time information about player competency states can be useful to support learning through hints and feedback as well as the dynamic matching of game difficulty level to player ability (e.g., providing more challenging problems for those with high levels of various skills). Regarding the example used in this report, the indicators linked to the agents of force and motion can serve as the basis for diagnoses. If a student created a lever that did not successfully solve a problem that could have been solved via a lever, for instance, the indicators would inform the most likely reason(s) why. That is, the lever may have failed given the wrong mass of an object that was used on one side of the lever, because the fulcrum was positioned inaccurately, and/or because the size/length of the lever was too short or too long. Those data (mass, position, and length) are calculated as part of the stealth assessment.

Specific future research in the area of stealth assessment includes working with middle-school teachers to embed *Newton's Playground* into the physical science curriculum. This will involve linking Newtonian physics formalizations (e.g., $F = ma$) to relevant *Newton's Playground* problems for instructional support. Teachers can also design their own levels in *Newton's Playground* to highlight physics concepts that could benefit from more hands-on experience and support. Additional scaling of the game includes adding more levels to game, especially interactions among Newton's laws of motion; creating more physics content, like principles of collision; examining predictive validity of the game relative to future science courses taken and grades received therein; using the indicators associated with the four agents of force and motion to infer misconceptions for diagnostic and support purposes; and expanding the platform of *Newton's Playground* from computer- to browser-based gameplay.

In the more distant future, we can foresee dynamic and unobtrusive assessments being used in classrooms as well as outside of school. The data from these assessments may be aggregated into rich and valid profiles of students, reducing (or removing) the need for the teach-stop-test model that has governed classroom instruction for too long. We can also imagine representations of "academic success" to go beyond letter grades. Just what does a C in algebra substantively mean?

We are excited that researchers are starting to use digital games for learning and assessment. We think stealth assessment is one way to maximize the positive impact that digital games can have on students.

Appendixes

Appendix 1: Full Physics Competency Model

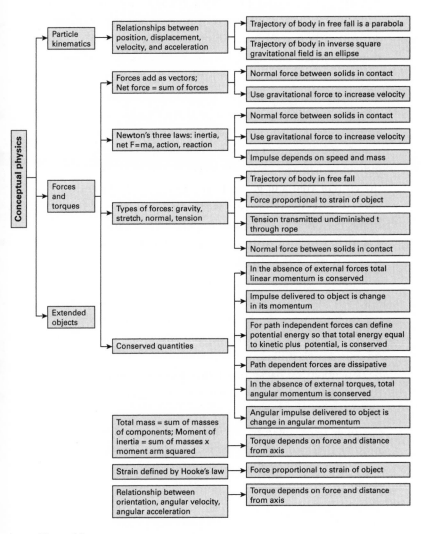

Figure 15

Appendix 2: External Measures to Validate Stealth Assessments

Performance-Based Measure of Persistence

We have developed a performance-based measure of persistence (PBMP) that measures how much effort people exert in difficult tasks (see Ventura, Shute, and Zhao 2012). The PBMP is administered online (in an Internet browser), and presents a variety of hard and easy problems (e.g., anagrams or picture comparison tasks) one at a time over a series of trials. Individuals type in their response and press the "guess" button. If the answer is wrong, the screen displays "incorrect" and the individual can try again (for up to 120 seconds). At any time the individual can also choose to select the "skip" button to leave the current trial and go on to the next one. If the individual guesses correctly, the person is told that he or she is correct. A trial is classified as "solved" if the person accurately completes the trial. A trial is classified as unsolved if the person skips the trial or is timed out after 120 seconds. We propose that persistence may motivate individuals to expend extra effort in solving hard problems outside their ability level. Specifically, the critical information in the PBMP that informs the assessment of persistence is *time spent on unsolved trials*. While the time spent on solved trials is likely a function of persistence as well, it may be dependent on background knowledge or ability in relation to the respective problem. Below are two screen captures: one of a "hard" anagram item (the correct answer is quisby), and one of a "hard" picture comparison task where five differences must be detected between the two pictures—four of which are fairly easy, and one of which is nearly impossible to find (see figure 16).

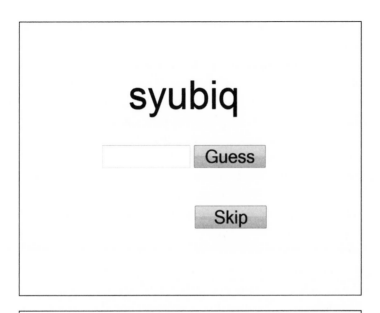

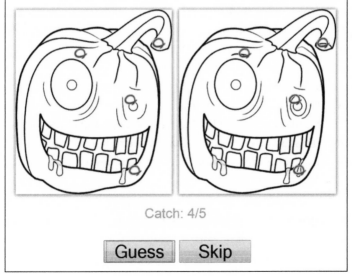

Figure 16

A PBMP has several potential advantages over traditional self-report measures. First, a PBMP can be seen as an implicit measure since no explicit questions are used that may cue the intentions of the assessment. This can mitigate the social desirability effect seen in self-report measures. Second, a PBMP can be claimed to have greater face validity than self-report measures. Self-report measures of persistence ask individuals how they act in difficult problems (e.g., "I never give up"), while performances-based assessments can actually measure behavior in real time on difficult problems. Finally, by nature of their implicit nature and face validity, a PBMP can offer a framework to assess learning of the target construct over time. For example, assessing the effectiveness of an intervention that tries to improve persistence might by compromised by using self-report measures (e.g., social desirability effects, different interpretation of items, lack of explicit knowledge of dispositional change). Alternatively, a PBMP may represent a more appropriate means to evaluate if persistence can be affected due to an experimental manipulation or lifestyle choice (e.g., playing video games).

Self-Report Items

We also plan to use a validated measure from the International Personality Item Pool (IPIP). Students will respond to the following 20 items on a 1–5 Likert scale (1 = strongly disagree to 5 = strong agree; asterisks denote items that are reverse keyed):

Persistence

1. I push myself very hard to succeed
2. I accomplish a lot of work
3. I have patience when it comes to difficult problems
4. I get easily frustrated on new problems*

5. I tend to give up easily*

6. I do more than what's expected of me

7. I tend to avoid difficult problems*

8. I put little time and effort into my work*

9. I enjoy a good challenge

10. I always try my hardest

Perfectionism

1. I dislike routine*

2. I pay attention to details

3. I continue until everything is perfect

4. I have an eye for detail

5. I want every detail taken care of

6. I dislike imperfect work

7. I want everything to add up perfectly

8. I detect mistakes

9. I demand quality

10. I prefer to just let things happen*

Performance-Based Measure of Creativity

Our external measures of creativity will include tests developed and validated by Wallach and Kogan as well as self-report items from the openness survey (also from the IPIP). For the former, we will follow a methodology developed by Wallach and Kogan for their widely used creativity test, which also has good psychometric properties (Wallach 1971). The instrument consists of three verbal tests (instances, alternate uses, and similarities) and two figural tests (abstract patterns and straight or curved lines) concerning the uses of common objects. We will use a version of their alternate uses test that asks questions such as "Tell me all the different ways you could use a book."

To ensure the reliability and validity of our version of the Wallach and Kogan creativity test, we will follow the scoring framework suggested by Paul Silvia and his colleagues (Silvia et al. 2008). That is, participants will be asked to circle their two most creative responses, and then human raters will judge their responses using a 5-point Likert scale (1 = not at all creative to 5 = highly creative). Based on the ratings provided by human raters, two creativity indexes will be used for the overall creativity score: the average creativity index (i.e., the sum of ratings across all responses divided by the number of responses), and the rating for the top two responses.

Self-Report Items

We plan to use a validated measure from the IPIP. Students will respond to the following 10 items on a 1–5 Likert scale (1 = strongly disagree to 5 = strong agree):

Openness

1. I like to think of new ideas
2. I enjoy art
3. I am excited by many different activities
4. I daydream a lot
5. I enjoy learning new things
6. I like to explore different solutions to problems
7. I have an active imagination
8. I like to be original
9. I try to be different from other students
10. I am curious about many different things

External Assessment of Conceptual Physics

We currently have a set of twenty-four items (i.e., twelve items in form A, and twelve isomorphic items in form B) that assess the competencies in our conceptual physics competency model. The test is divided among the four main agents of force and motion. Within each section, different facets of the physics principles are assessed. Items are either multiple choice or constructed response. For constructed response items (like the one shown in figure 17), our rubric will consist of an optimal trajectory surrounded by an area comprising a "correct response."

For our multiple-choice items, the format is generally the same—where a problem is presented, along with a graphic that

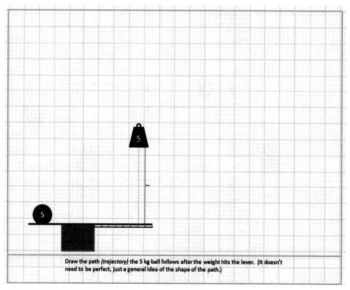

Draw the path *(trajectory)* the 5 kg ball follows after the weight hits the lever. (It doesn't need to be perfect, just a general idea of the shape of the path.)

Clear   Next

Figure 17

consists of two options (A and B). The student has to decide whether the graphic depicted in A is correct, B is correct, both A and B are equal, or the answer isn't known. See the example shown in figure 18.

The items were created based on Hewitt's 2009 textbook (*Conceptual Physics*, eleventh ed.), and then reviewed and edited by our physicist working on the project (Dr. Donald Franceschetti, University of Memphis). We also plan to use items from the Force Concept Inventory (Hestenes, Wells, and Swackhamer 1992) to measure for the transfer of physics principles. For example, we expect that playing *Newton's Playground* will result in the conceptual understanding of various object collisions (e.g., moment of inertia) not explicitly observed in *Newton's Playground*.

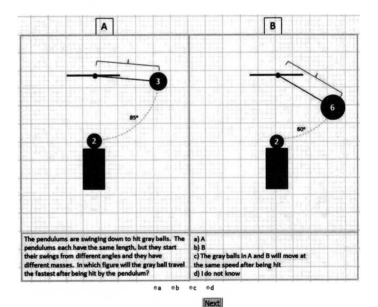

Figure 18

References

Abe, J.A.A. 2005. The Predictive Validity of the Five-Factor Model of Personality with Preschool Age Children: A Nine-Year Follow-up Study. *Journal of Research in Personality* 39:423–442.

Almond, R. G., and R. J. Mislevy. 1999. Graphical Models and Computerized Adaptive Testing. *Applied Psychological Measurement* 23:223–237.

Amabile, T. M. 1983. The Social Psychology of Creativity: A Componential Conceptualization. *Journal of Personality and Social Psychology* 45, no. 2:357–376. doi:10.1037/0022-3514.45.2.357.

Armstrong, A., and H. Georgas. 2006. Using Interactive Technology to Teach Information Literacy Concepts to Undergraduate Students. *Reference Services Review* 34, no. 4:491–497.

Barab, S. A., T. Dodge, A. Ingram-Goble, P. Pettyjohn, K. Peppler, C. Volk, and M. Solomou. 2010. Pedagogical Dramas and Transformational Play: Narratively Rich Games for Learning. *Mind, Culture, and Activity* 17, no. 3:235–264.

Barab, S. A., M. Gresalfi, and A. Ingram-Goble. 2010. Transformational Play. *Educational Researcher* 39, no. 7:525–536. doi:10.3102/0013189x10386593.

Barab, S. A., S. Zuiker, S. Warren, D. Hickey, A. Ingram-Goble, E.-J. Kwon, I. Kouper, and S. C. Herring. 2007. Situationally Embodied Curriculum: Relating Formalisms and Contexts. *Science Education* 91, no. 5:750–782.

Berzonsky, M. D., and C. Sullivan. 1992. Social-Cognitive Aspects of Identity Style. *Journal of Adolescent Research* 7, no. 2:140–155. doi:10.1177/074355489272002.

Bridgeland, J. M., J. J. DiIulio Jr., and K. B. Morison. 2006. *The Silent Epidemic: Perspectives of High School Dropouts*. Washington, DC: Civic Enterprises and Peter D. Hart Research Associates.

Chory, R. M., and A. K. Goodboy. 2011. Is Basic Personality Related to Violent and Non-violent Video Game Play and Preferences? *Cyberpsychology, Behavior, and Social Networking* 14, no. 4:191–198.

Coller, B. D., and M. J. Scott. 2009. Effectiveness of Using a Video Game to Teach a Course in Mechanical Engineering. *Computers and Education* 53, no. 3:900–912. doi:10.1016/j.compedu.2009.05.012.

Costa, P. T., Jr., and R. R. McCrae. 1992. *Revised NEO Personality and Five Factor Inventory Professional Manual*. Odesa, FL: Psychological Assessment Resources.

Crouch, C. A., and E. Mazur. 2001. Peer Instruction: Ten Years of Experience and Results. *American Journal of Physics* 69:970–977.

Csikszentmihalyi, M. 1990. The Domain of Creativity. In *Theories of Creativity*, ed. M. A. Runco and R. S. Albert, 190–212. Newbury Park, CA: Sage.

Dede, C. 2005. Planning for Neomillennial Learning Styles. *EDUCAUSE Quarterly* 28, no. 1:7–12.

De Fruyt, F., and I. Mervielde. 1996. Personality and Interests as Predictors of Educational Streaming and Achievement. *European Journal of Personality* 10:405–425.

DeRouin-Jessen, R. 2008. Game On: The Impact of Game Features in Computer-Based Training. PhD diss., University of Central Florida, Orlando.

Dewett, T. 2007. Linking Intrinsic Motivation, Risk Taking, and Employee Creativity in an R&D Environment. *R & D Management* 37, no. 3:197–208. doi:10.1111/j.1467-9310.2007.00469.x.

DiCerbo, K. E., and J. T. Behrens. 2012. Implications of the Digital Ocean on Current and Future Assessment. In *Computers and Their Impact on State Assessment: Recent History and Predictions for the Future*, ed. R. Lissitz and H. Jiao, 273–306. Charlotte, NC: Information Age Publishing.

diSessa, A. A. 1982. Unlearning Aristotelian Physics: A Study of Knowledge-Based Learning. *Cognitive Science* 6:37–75.

Eisenberg, R. 1992. Learned Industriousness. *Psychological Review* 99, no. 2:248–267.

Eisenberg, R., and J. M. Leonard. 1980. Effects of Conceptual Task Difficulty on Generalized Persistence. *American Journal of Psychology* 95, no. 2:285–298.

Entertainment Software Association. 2011. Essential Facts about the Computer and Video Game Industry. http://www.theesa.com/facts/pdfs/ESA_EF_2011.pdf.

Escobar-Chaves, S. L., and C. A. Anderson. 2008. Media and Risky Behaviors. *Future of Children* 18, no. 1:147–180.

Feist, G. J. 1999. The Influence of Personality on Artistic and Scientific Creativity. In *Handbook of Creativity*, ed. R. J. Sternberg, 273–296. New York: Cambridge University Press.

Ferguson, C. J., and A. Garza. 2011. Call of (Civic) Duty: Action Games and Civic Behavior in a Large Sample of Youth. *Computers in Human Behavior* 27:770–775.

Feynman, R. P. 1964. *The Character of Physical Law*. Ithaca, NY: Cornell University Press.

Feynman, R. P., R. B. Leighton, and M. Sands. 1964. *The Feynman Lectures in Physics*. Boston: Addison-Wesley.

Friedman, T. 2005. *The World Is Flat: A Brief History of the Globalized World in the Twenty-First Century.* London: Allen Lane.

Fullerton, T. 2008. *Game Design Workshop: A Playcentric Approach to Creating Innovative Games.* Burlington, MA: Elsevier.

Gee, J. P. 2003. *What Video Games Have to Teach Us about Learning and Literacy.* New York: Palgrave Macmillan.

Gee, J. P., G. A. Hull, and C. Lankshear. 1996. *The New Work Order: Behind the Language of the New Capitalism.* St Leonards, Australia: Allen and Unwin.

Glover, J. A. 1977. Risky Shift and Creativity. *Social Behavior and Personality: An International Journal* 5, no. 2:317.

Glover, J. A., and F. Sautter. 1977. Relation of Four Components of Creativity to Risk-Taking Preferences. *Psychological Reports* 41, no. 1:227–230.

Green, C. S., and D. Bavelier. 2007. Action-Video-Game Experience Alters the Spatial Resolution of Vision. *Psychological Science* 18, no. 1:88–94.

Gronhaug, K., and G. Kaufman, eds. 1988. *Innovation: A Cross-Disciplinary Perspective.* Oslo: Norwegian Universities Press.

Guilford, J. P. 1956. The Structure of Intellect. *Psychological Bulletin* 53:267–293. doi:10.1037/h0040755.

Hake, R. R. 1998. Interactive Engagement vs. Traditional Methods in Mechanics Instruction. *American Journal of Physics* 66, no. 1:64–74.

Halloun, I. 1996. Schematic Modeling for Meaningful Learning of Physics. *Journal of Research in Science Teaching* 33:407–431.

Halloun, I., and D. Hestenes. 1985. Initial Knowledge State of College Physics Students. *American Journal of Physics* 53:1043–1055.

Hattie, J., and H. Timperley. 2007. The Power of Feedback. *Review of Educational Research* 77, no. 1:81–112.

Hestenes, D., and M. Wells. 1992. A Mechanics Baseline Test. *Physics Teacher* 30, no. 3:159–167.

Hestenes, D., M. Wells, and G. Swackhamer. 1992. Force Concept Inventory. *Physics Teacher* 30:141–151.

Hewitt, P. G. 2009. *Conceptual Physics*. 11th ed. San Francisco: Pearson Education.

Hunsaker, S. L., and C. M. Callahan. 1995. Creativity and Giftedness: Published Instrument Uses and Abuses. *Gifted Child Quarterly* 39, no. 2:110–114. doi:10.1177/001698629503900207.

Ito, M., S. Baumer, M. Bittanti. d. boyd, R. Cody, B. Herr-Stephenson, H. A. Horst, P. G. Lange, D. Mahendran, K. Martinez, C. J. Pascoe, D. Perkel, L. Robinson, C. Sims, and L. Tripp. 2010. *Hanging Out, Messing Around, and Geeking Out: Kids Living and Learning with New Media*. Cambridge, MA: MIT Press.

Kato, P. M., S. W. Cole, A. S. Bradlyn, and B. H. Pollock. 2008. A Video Game Improves Behavioral Outcomes in Adolescents and Young Adults with Cancer: A Randomized Trial. *Pediatrics* 122, no. 2:305–317.

Kaufman, G. 2003. What to Measure? A New Look at the Concept of Creativity. *Scandinavian Journal of Educational Research* 47, no. 3:235–251.

Kaufman, J. C., and R. J. Sternberg. 2007. Resource Review: Creativity. *Change* 39, no. 4:55–58.

Kellaghan, T., and G. F. Madaus. 1991. National Testing: Lessons for America from Europe. *Educational Leadership* 49, no. 3:87–93.

Kliebard, H. 1987. *The Struggle for the American Curriculum, 1893–1958*. New York: Routledge and Kegan Paul.

Kogan, N., and M. A. Wallach. 1964. *Risk Taking: A Study in Cognition and Personality*. Oxford: Holt, Rinehart and Winston.

Korbin, J. L., B. F. Patterson, E. J. Shaw, K. D. Mattern, and S. M. Barbuti. 2008. Validity of the SAT for Predicting First-Year College Grade Point Average (Research Report No. 2008–5). New York: College Board.

Lanyon, R. I., and L. D. Goodstein. 1997. *Personality Assessment.* 3rd ed. New York: Wiley.

Lenhart, A., J. Kahne, E. Middaugh, A. R. Macgill, C. Evans, and J. Vitak. 2008. *Teens' Gaming Experiences Are Diverse and Include Significant Social Interaction and Civic Engagement.* Washington, DC: Pew Internet and American Life Project.

Lim, C. P., D. Nonis, and J. Hedberg. 2006. Gaming in a 3-D Multiuser Virtual Environment: Engaging Students in Science Lessons. *British Journal of Educational Technology* 37, no. 2:211–231.

Lubart, T. I. 1994. Creativity. In *Thinking and Problem Solving*, ed. R. J. Sternberg, 289–332. New York: Academic.

MacCann, C., A. L. Duckworth, and R. D. Roberts. 2009. Empirical Identification of the Major Facets of Conscientiousness. *Learning and Individual Differences* 19:451–458.

MacCrimmon, K. R., and D. A. Wehrung. 1990. Characteristics of Risk Taking Executives. *Management Science* 36, no. 4:422–435.

Madaus, G., and L. O'Dwyer. 1999. A Short History of Performance Assessment. *Phi Delta Kappan* 80, no. 9:688–695.

Madjar, N., E. Greenberg, and Z. Chen. 2011. Factors for Radical Creativity, Incremental Creativity, and Routine, Noncreative Performance. *Journal of Applied Psychology* 96, no. 4:730–743. doi:10.1037/a0022416.

McCrae, R. R. 1987. Creativity, Divergent Thinking, and Openness to Experience. *Journal of Personality and Social Psychology* 52, no. 6:1258–1265. doi:10.1037/0022-3514.52.6.1258.

McCrae, R. R. 1996. Social Consequences of Experiential Openness. *Psychological Bulletin* 120:323–337.

McDermott, L. 1993. How We Teach and How Students Learn: A Mismatch? *American Journal of Physics* 61:295–298.

Messick, S. 1989. Validity. In *Educational Measurement*, ed. R. L. Linn, 13–104. 3rd ed. New York: Macmillan.

Messick, S. 1994. The Interplay of Evidence and Consequences in the Validation of Performance Assessments. *Educational Researcher* 23, no. 2:13–23.

Mislevy, R. J., and G. D. Haertel. 2006. Implications of Evidence-Centered Design for Educational Testing. *Educational Measurement: Issues and Practice* 25, no. 4:6–20.

Mislevy, R. J., L. S. Steinberg, and R. G. Almond. 2003. On the Structure of Educational Assessments. *Measurement: Interdisciplinary Research and Perspectives* 1, no. 1:3–62.

NCLB: No Child Left Behind Act of 2001. 2002. Pub. L. No. 107–110, 115 Stat. 1425.

Nijstad, B. A., C.K.W. De Dreu, E. F. Rietzschel, and M. Baas. 2010. The Dual Pathway to Creativity Model: Creative Ideation as a Function of Flexibility and Persistence. *European Review of Social Psychology* 21, no. 1:34–77. doi:10.1080/10463281003765323.

Nitko, A. J. 1980. Distinguishing the Many Varieties of Criterion-Referenced Tests. *Review of Educational Research* 50:461–485.

Noftle, E. E., and R. W. Robins. 2007. Personality Predictors of Academic Outcomes: Big Five Correlates of GPA and SAT Scores. *Journal of Personality and Social Psychology* 93:116–130.

O'Connor, M., and S. Paunonen. 2007. Big Five Personality Predictors of Post-secondary Academic Performance. *Personality and Individual Differences* 43:971–990.

Okuda, S. M., M. A. Runco, and D. E. Berger. 1991. Creativity and the Finding and Solving of Real-world Problems. *Journal of Psychoeducational Assessment* 9:45–53.

Paulhaus, D. L. 1991. Measurement and Control of Response Bias. In *Measures of Personality and Social Psychological Attitudes: Volume 1 of Measures of Social Psychological Attitudes*, ed. J. P. Robinson, P. R. Shaver, and L. S. Wrightsman, 17–59. San Diego: Academic Press.

Partnership for the 21st Century. 2012. http://www.p21.org.

Peabody, D., and B. De Raad. 2002. The Substantive Nature of Psycho-lexical Personality Factors: A Comparison across Languages. *Journal of Personality and Social Psychology* 83:983–997.

Pellegrino, J. W., N. Chudowsky, and R. Glaser. 2001. *Knowing What Students Know: The Science and Design of Educational Assessment.* Washington, DC: National Academy Press.

Perugini, M., and M. Gallucci. 1997. A Hierarchical Faceted Model of the Big Five. *European Journal of Personality* 11:279–301.

Poropat, A. E. 2009. A Meta-analysis of the Five-Factor Model of Personality and Academic Performance. *Psychological Bulletin* 135:322–338.

Quellmalz, E. S., M. J. Timms, B. C. Buckley, M. Silberglitt, and D. Brenner. 2012. SimScientists: Measurement in Simulation-Based Science Assessments. Unpublished manuscript.

Robbins, S. B., K. Lauver, H. Le, D. Davis, R. Langley, and A. Carlstrom. 2004. Do Psychosocial and Study Skill Factors Predict College Outcomes? A Meta-analysis. *Psychological Bulletin* 130:261–288.

Roberts, B. W., T. Bogg, K. E. Walton, O. S. Chernyshenko, and S. E. Stark. 2004. A Lexical Investigation of the Lower-Order Structure of Conscientiousness. *Journal of Research in Personality* 38:164–178.

Roberts, B. W., O. S. Chernyshenko, S. E. Stark, and L. R. Goldberg. 2005. The Structure of Conscientiousness: An Empirical Investigation Based on Seven Major Personality Questionnaires. *Personnel Psychology* 58:103–139.

Roberts, B. W., N. Kuncel, R. N. Shiner, A. Caspi, and L. R. Goldberg. 2007. The Power of Personality: The Comparative Validity of Personality Traits, Socio-economic Status, and Cognitive Ability for Predicting Important Life Outcomes. *Perspectives on Psychological Science* 2:313–345.

Roberts, B. W., K. Walton, and W. Viechtbauer. 2006. Patterns of Mean-Level Change in Personality Traits across the Life Course: A Meta-analysis of Longitudinal Studies. *Psychological Bulletin* 132:1–25.

Runco, M. A. 1986. Flexibility and Originality in Children's Divergent Thinking. *Journal of Psychology* 120, no. 4:345.

Runco, M. A. 1992. Children's Divergent Thinking and Creative Ideation. *Developmental Review* 12:233–264.

Runco, M. A., ed. 1997. *The Creativity Research Handbook*. Vol. 1. Cresskill, NJ: Hampton Press.

Runco, M. A., ed. 2002. *The Creativity Research Handbook*. Vol. 2. Cresskill, NJ: Hampton Press.

Runco, M. A. 2008. Commentary: Divergent Thinking Is Not Synonymous with Creativity. *Psychology of Aesthetics, Creativity, and the Arts* 2:93–96.

Runco, M. A., and S. R. Pritzker. 1999. *Encyclopedia of Creativity*. Vol. 1. New York: Academic Press.

Rupp, A. A., M. Gushta, R. J. Mislevy, and D. W. Shaffer. 2010. Evidence-Centered Design of Epistemic Games: Measurement Principles for Complex Learning Environments. *Journal of Technology, Learning, and Assessment* 8, no. 4. http://ejournals.bc.edu/ojs/index.php/jtla/article/view/1623.

Salen, K., and E. Zimmerman. 2005. Game Design and Meaningful Play. In *Handbook of Computer Game Studies*, ed. J. Raessens and J. Goldstein, 59–80. Cambridge, MA: MIT Press.

Saucier, G., and F. Ostendorf. 1999. Hierarchical Subcomponents of the Big Five Personality Factors: A Cross-Language Replication. *Journal of Personality and Social Psychology* 76, no. 4:613–627.

Schmidt, F. L., and J. E. Hunter. 1998. The Validity and Utility of Selection Methods in Personnel Psychology. *Psychological Bulletin* 124:262–274.

Schmitt, N. 1994. Method Bias: The Importance of Theory and Measurement. *Journal of Organizational Behavior* 15:393–398.

Shaw, R., A. Grayson, and V. Lewis. 2005. Inhibition, ADHD, and Computer Games: The Inhibitory Performance of Children with ADHD on

Computerized Tasks and Games. *Journal of Attention Disorders* 8, no. 4:160–168. doi:10.1177/1087054705278771.

Shepard, L. A. 1991. Will National Tests Improve Student Learning? *Phi Delta Kappan* 72:232–238. http://ww.cse.ucla.edu/products/Reports/TECH342.pdf.

Shiner, R. L., A. S. Masten, and J. M. Roberts. 2003. Childhood Personality Foreshadows Adult Personality and Life Outcomes Two Decades Later. *Journal of Personality* 71:1145–1170.

Shute, V. J. 2007. Tensions, Trends, Tools, and Technologies: Time for an Educational Sea Change. In *The Future of Assessment: Shaping Teaching and Learning*, ed. C. A. Dwyer, 139–187. New York: Erlbaum.

Shute, V. J. 2008. Focus on Formative Feedback. *Review of Educational Research* 78, no. 1:153–189.

Shute, V. J. 2009. Simply Assessment. *International Journal of Learning and Media* 1, no. 2:1–11.

Shute, V. J. 2011. Stealth Assessment in Computer-Based Games to Support Learning. In *Computer Games and Instruction*, ed. S. Tobias and J. D. Fletcher, 503–524. Charlotte, NC: Information Age Publishers.

Shute, V. J., E. G. Hansen, and R. G. Almond. 2008. You Can't Fatten a Hog by Weighing It—Or Can You? Evaluating an Assessment for Learning System Called ACED. *International Journal of Artificial Intelligence in Education* 18, no. 4:289–316.

Shute, V. J., and F. Ke. 2012. Games, Learning, and Assessment. In *Assessment in Game-Based Learning: Foundations, Innovations, and Perspectives*, ed. D. Ifenthaler, D. Eseryel, and X. Ge, 43–58. New York: Springer.

Shute, V. J., and Y. J. Kim. 2011. Does Playing the *World of Goo* Facilitate Learning? In *Design Research on Learning and Thinking in Educational Settings: Enhancing Intellectual Growth and Functioning*, ed. D. Y. Dai, 359–387. New York: Routledge.

Shute, V. J., Y. J. Kim, and R. Razzouk. 2010. ECD for Dummies. http://myweb.fsu.edu/vshute/ECD%20for%20Dummies/ECD%20for%20Dummies.swf.

Shute, V. J., R. Levy, R. Baker, D. Zapata, and J. Beck. 2009. Assessment and Learning in Intelligent Educational Systems: A Peek into the Future. In *Proceedings of the Artificial Intelligence and Education (AIED '09) Workshop on Intelligent Educational Games*, ed. S. D. Craig and D. Dicheva, 99–109. Brighton, UK.

Shute, V. J., I. Masduki, and O. Donmez. 2010. Conceptual Framework for Modeling, Assessing, and Supporting Competencies within Game Environments. *Technology, Instruction, Cognition, and Learning* 8, no. 2:137–161.

Shute, V. J., and R. Torres. 2012. Where Streams Converge: Using Evidence-Centered Design to Assess Quest to Learn. In *Technology-Based Assessments for Twenty-First-Century Skills: Theoretical and Practical Implications from Modern Research*, ed. M. Mayrath, J. Clarke-Midura, and D. H. Robinson, 91–124. Charlotte, NC: Information Age Publishing.

Shute, V. J., M. Ventura, M. I. Bauer, and D. Zapata-Rivera. 2009. Melding the Power of Serious Games and Embedded Assessment to Monitor and Foster Learning: Flow and Grow. In *Serious Games: Mechanisms and Effects*, ed. U. Ritterfeld, M. Cody, and P. Vorderer, 295–321. Mahwah, NJ: Routledge, Taylor and Francis.

Silvia, P. J., B. P. Winterstein, J. T. Willse, C. M. Barona, J. T. Cram, K. I. Hess, J. L. Martinez, and C. A. Richard. 2008. Assessing Creativity with Divergent Thinking Tasks: Exploring the Reliability and Validity of New Subjective Scoring Methods. *Psychology of Aesthetics, Creativity, and the Arts* 2, no. 2:68–85.

Skoric, M. M., L.L.C. Teo, and R. L. Neo. 2009. Children and Video Games: Addiction, Engagement, and Scholastic Achievement. *Cyberpsychology and Behavior* 12, no. 5:567–572.

Snow, C. E., and J. Jones. 2001. Making a Silk Purse. *Education Week Commentary* 20, no. 32:60.

Sternberg, R. J., ed. 1988. *The Nature of Creativity*. Cambridge: Cambridge University Press.

Sternberg, R. J., ed. 1999. *Handbook of Creativity*. Cambridge: Cambridge University Press.

Sternberg, R. J. 2006. The Nature of Creativity. *Creativity Research Journal* 18, no. 1:87–98.

Sternberg, R. J., and T. I. Lubart. 1992. Buy Low and Sell High: An Investment Approach to Creativity. *Current Directions in Psychological Science* 1, no. 1:1–5.

Swann, W.F.G. 1950. The Teaching of Physics. *American Journal of Physics* 19, no. 2:182–187.

Terman, L. M. 1916. *The Measurement of Intelligence.* Cambridge, MA: Riverside Press.

Tobias, S., and J. D. Fletcher, eds. 2011. *Computer Games and Instruction.* Charlotte, NC: Information Age Publishers.

Tobias, S., J. D. Fletcher, D. Y. Dai, and A. P. Wind. 2011. Review of Research on Computer Games. In *Computer Games and Instruction*, ed. S. Tobias and J. D. Fletcher, 127–222. Charlotte, NC: Information Age.

Torrance, E. P. 1974. *Torrance Tests of Creative Thinking: Norms-Technical Manual.* Lexington, MA: Ginn and Company.

Trapmann, S., B. Hell, J. W. Hirn, and H. Schuler. 2007. Meta-analysis of the Relationship between the Big Five and Academic Success at University. *Journal of Psychology* 215:132–151.

Trilling, B., and C. Fadel. 2009. *Twenty-First-Century Skills: Learning for Life in Our Times.* San Francisco: Jossey-Bass.

Ventura, M., V. J. Shute, and Y. J. Kim. 2012. Video Gameplay, Personality, and Academic Performance. *Computers and Education* 58:1260–1266.

Ventura, M., V. J. Shute, and W. Zhao. 2012. The Relationship between Video Game Use and a Performance-Based Measure of Persistence. *Computers and Education* 60:52–58.

Vygotsky, L. 1978. *Mind in Society: The Development of Higher Psychological Processes.* Cambridge, MA: Harvard University Press.

Walberg, H. J., and W. E. Stariha. 1992. Productive Human Capital: Learning, Creativity, and Eminence. *Creativity Research Journal* 5:323–340.

Wallach, M. A. 1971. *The Intelligence/Creativity Distinction.* New York: General Learning Press.

Wallach, M. A., and N. Kogan. 1965. *Modes of Thinking in Young Children.* New York: Holt, Rinehart and Winston.

Weisberg, R. W. 1993. *Creativity: Beyond the Genius.* New York: Freeman.

White, B. Y. 1994. Designing Computer Games to Help Physics Students Understand Newton's Laws of Motion. *Cognition and Instruction* 1, no. 1:69–108.

Wilson, K. A., W. Bedwell, E. H. Lazzara, E. Salas, C. S. Burke, J. Estock, and C. Conkey. 2009. Relationships between Game Attributes and Learning Outcomes: Review and Research Proposals. *Simulation and Gaming* 40 no. 2:217–266.

Witt, E. A., A. J. Massman, and L. A. Jackson. 2011. Trends in Youth's Videogame Playing, Overall Computer Use, and Communication Technology Use: The Impact of Self-esteem and the Big Five Personality Factors. *Computers in Human Behavior* 27, no. 2:763–769.

Zapata-Rivera, D., and M. Bauer. 2011. Exploring the Role of Games in Educational Assessment. In *Technology-Based Assessments for Twenty-First-Century Skills: Theoretical and Practical Implications from Modern Research*, ed. M. Mayrath, J. Clarke-Midura, D. Robinson, and G. Shraw, 147–169. Charlotte, NC: Information Age Publishing.